OBDII DIAGNOSTIC > SECRETS REVEALED

Written by experts in the field

Published by:

Kotzig Publishing

2505 NW Boca Raton Blvd

Suite 205

Boca Raton, FL 33431

send your comments to:

obdii@kotzigpublishing.com

Printed in Slovakia

ISBN 0-9715411-4-0

Disclaimer. This publication is being sold with no warranties of any kind, express or implied. This publication is intended for educational and informational purposes only, and should not be relied upon as a "how-to" guide in performing repairs, maintenance or modification of your vehicle. This publication is not endorsed by the manufacturer of your vehicle nor DiabloSport or ToolRama, and there is no affiliation between the publisher and author and the manufacturer of your vehicle.

Modifications performed on your vehicle may limit or void your rights under any warranty provided by the manufacturer of your vehicle, and neither the publisher nor the author of this publication assume any responsibility in such event. Any warranty not provided herein, and any remedy which, but for this provision, might arise by implication or operation of law, is hereby excluded and disclaimed. The implied warranties of merchantability and of fitness for any particular purpose herein are expressly disclaimed.

No Liability For Damages, Injuries Or Incidental, Special Or Consequential Damages. Under no circumstances shall the publisher, author, or contributing author of this publication, or any other related party, be liable to purchaser or any other person for any damage to your vehicle, loss of use of your vehicle, or for personal injuries suffered by any person, or for any incidental, special or consequential damages, whether arising out of negligence, breach of warranty, breach of contract, or otherwise.

State Law. Some states do not allow limitations of implied warranties, or the exclusion or limitation of incidental, special or consequential damages, so the above limitations may not apply to you. In such states, liability shall be limited to the greatest extent permitted by applicable law.

Warning. Although the publisher and author have made every effort to ensure accuracy of all information contained in this publication, the reader should not attempt any modifications of his or her vehicle or computer codes without the assistance of a qualified and trained technician. Always obey state and federal laws, follow manufacturers' operating instructions and observe safety precautions.
All information contained in the publication is for educational and informational purposes only and is not intended as a "how-to" guide.

Note. All trademarks are property of their individual owners.

To the Memory of Chris Barone

and all the other experienced technicians,

who took the time to teach us

to become technicians,

rather than just mechanics.

Contents

ENTER

Preface

ENTER

If you had told me six months ago, that I would be writing a book, I would have laughed you out of the room. Now, it has become a reality. The motto I have been using for some years now is "never say never" and "always keep an open mind." The point I am trying to make is that twenty-three years ago I started to work at a lawnmower shop. My father was tired of my taking apart the lawnmower, vacuum cleaner, clothes washer, etc., so he took me down to the local lawnmower shop and got me a job.

After four years of doing that, I decided it was time to move on and start my future at a local Jaguar dealer. I wanted to start at the top, and if I got a job at the bottom... well, at least I tried. There I was trained by a tech named Chris Barone. Chris was by no means a gentleman, but he was a great tech. He taught me by giving me as much junk work as I could handle. He constantly ridiculed and tortured me. I didn't like him at all. For a while, I thought of quitting, but I hung in there. It wasn't until years later that I understood what he was doing. Chris was getting me not to rely on him for help. He wanted me to get so angry that I would figure out the problem, even if it killed me, to show him that I could do it. That is what I did. I figured it out, and relied on myself. Chris taught me in a short time to be a self-sufficient technician. He also taught me a trade that is now dying out. How many guys can set up multiple carbs on an Aston Martin or Ferrari? I still have my mercury tree.

Anyway, I worked on Jaguars, Aston Martins, Ferraris and Rolls for some years, until I found (at least in my mind) the Mecca of cars: Mercedes Benz. What a wonderful car to repair! I stayed at a Mercedes dealer for some years and then moved on to an independent shop. That is where I really learned to be a true technician. At dealers you get the training, but you become lazy. You have a vast parts department to swap parts, instead of figuring out what the problem is.

In the automotive field you get paid by the job. If a job pays an hour, and it takes you thirty minutes to do, you made money. If you did the job in two hours, you lost money. So, if you have a vast parts department, and you have the opportunity to swap a part, well, you do it, and still get paid for diagnosis and replacing the part. At an independent shop, you don't have that luxury. If you order a part, you better be darn sure that the part is bad - if you are wrong, you pay for that part! This forces you to diagnose correctly the first time, which makes you a better technician.

After I was at the independent for close to 6 years, I received a call from ToolRama, a company that designs and manufactures scan tools. They wanted to know if I wanted to work for them. At first I was a little skeptical, but later I jumped at the idea. I got to get out of turning wrenches, but stay in the repair field. ToolRama wanted someone for support of the scan tool. Today, I am a key member of the scan tool group.

If you had told me four years ago that I would be helping to develop a scan tool, I would have laughed you out of the room. Now you see that the mottos "never say never" and "always keep an open mind" have brought me this far.

The future... is yet to come.

Ackowledgments

I would like to thank
a few people that have helped
to make this book a reality...

ENTER

First my wife Justina. She has been very patient with me throughout the creation of this book. Justina has also added a cooking chapter (the publisher was pushing for this...). I think she is a great cook, with some recipes that not many people know about. At first she was a little skeptical about doing it, but, in the end, she went far beyond what I ever expected.

Next, Ivan Kotzig, the publisher, and my boss. He saw something in me and gave me a chance. I hope I have not let him down. Ivan is a long time designer of automotive electronics, and has a Master's degree in Engineering. He has dedicated his career to creating many inventive electronic circuits for the car repair industry.

Now, I would like to thank the people behind the scenes. Without them, none of this would happen...

Ruben Arnejo-Fernandez, well, he is a one of a kind person. He is the genius behind the graphic design. Ruben was born in Urugay, educated in classic drawing, painting, and photography. He is responsible for making my words come to life.
(Ruben came up with the idea to put an owl on the Secrets Revealed stamp. He refused to remove it – so we left it there to make him happy!).

Susan McCabe, she is the one that has spent countless hours reading and correcting my horrible grammar. She has had to read this book, even though she has no interest in this subject.

For the people that helped me get to where I am today...
First there are my parents, Pat and Alex David. They saw the ability in me, and tried to make me see it as well. At age 16, they tried to get me to go into the computer field, but I wanted to be a mechanic, and thought that those computer geeks were... well, let's not go there, since today I am one of those computer geeks. They saw the future and I ignored them. I finally got there, but the hard way.

John Anderson hired me at the request of my father. Thank you for giving me a chance.

Billy Moore is the one that helped me to get the job at the Jag dealer and has always been there for me.

Chris Barone, I hated you at first, but I love you now. You helped me far more than you ever expected.

Paul DeGuglimo, a friend and the best darn parts guy in the world.

Frank Ortiz gave me a chance when no one else would. He was the best man at my wedding and continues to be there for me.

Lastly, I would like to thank Gerhard and Michael Assenmacher. Without them, there would be no scan tool. Without a scan tool, I would still be turning wrenches.

I hope you enjoy this book. If you don't like it, look up all the people I have mentioned and complain to them, since they are responsible for my being in a position to write this book.

Today, I am very glad I hung in there!

Introduction

Now you too can understand the details of diagnosing OBDII vehicles!

This book is both an OBDII bible and the first complete guide to diagnosing modern automobiles; both for absolute beginners and advanced technicians alike.
If you want to understand the principles of electronic diagnostics, this book is for you!
Detailed descriptions will answer all of your questions and more.

It is the most comprehensive collection of information on OBDII ever printed.
Now, the secrets of experienced technicians can be yours!

Author Peter David, longtime technician at many fine dealerships, presents this interesting subject in layman's terms.

Peter has been an automotive technician for nineteen years, with expertise in European and Exotic vehicles. He spent over twelve years at various dealerships such as Aston Martin, Jaguar and Mercedes-Benz, and spent an additional seven years as Shop Foreman for an independent repair facility.

Several years ago he joined the team of ToolRama as Chief Researcher and has devoted his experience to the development of one of the leading diagnostic tools in the automotive world.

ENTER

About this book

What is in this book

This book is an OBDII bible.

As such, it doesn't have

to be read from cover-to-cover,

regardless of the author's dream

that you do so!

While other books can be very boring, we bring OBDII to life. We will also show you why OBDII is important to our lives and our health.

What is in this book

We recommend that you read this chapter, so you will have an idea which parts of the book you really need. Most novices will likely need the whole book; a master auto technician might not. So, dig in and enjoy!

On-Board Diagnostic II (OBDII) is a standard that almost everyone today has seen, heard of or used. Most people don't even know that OBDII is influencing their vehicle. There are millions of vehicles on the road with OBDII, influencing all our lives with cleaner air. It may very well be today's least known, but most used automotive concept.

Ever wondered what makes an automotive engine run? In the first chapter, **Basic Concepts of an Engine**, you will get an understanding of how an engine really works. If you are not familiar with the basics of an engine, OBDII will not make much sense. We are not going to get into complex issues, just simple facts.

The **Basics of Powertrain Control Module (PCM)** chapter is an introduction to the layman's world of automotive electronics. It should satisfy your curiosity about what is inside the computer box controlling your entire engine and OBDII diagnostics.

The next part of this book is about the **History of OBDII**. Before you can really understand what something is, you must first understand why and where it came from. The origin of OBDII dates all the way back to 1955. Since then, it has been a rocky road from the clean air activists to the independent automotive repair shops. Many people and organizations fought hard to bring about what we have today. In addition, in the history chapter, we will discuss all the organizations that helped OBDII become what it is today.

After the history, we will explain what makes up OBDII. We call it the **Anatomy of OBDII**. Think of it as a human body: OBDII has many parts, and each part is crucial to its existence. We will dissect each part of OBDII, including the diagnostic connector; the protocols and their classifications; the break down of the fault codes and the terminology of each component.

Next, on to the protocols. With an understanding of the parts of OBDII, the next logical step is what language OBDII speaks. You will learn in the chapter **Understanding OBDII Protocols**, how the control units communicate. In addition to the control units, a scan tool must use these protocols as well. You will be enlightened by learning about the pulse widths, modulations, high-speed communications and all four possible bus interfaces in OBDII compatible vehicles.

The next section discusses what you use to diagnose OBDII equipped vehicles. This is commonly known as the **OBDII Scan Tool**. We will talk about what types of tools are out there. You could have a handheld stand-alone scanner, or a PDA type, or even just software that you can install on your laptop. After you know the hardware, you must know what its capabilities are.

Now that you have decided which tool to use, how are you going to use it? In the chapter **Diagnostics with a Scan Tool** we will show you how to check which system is monitored; how to retrieve faults; how to check what the engine was doing when the fault happened; real time data values; and oxygen sensor monitoring.

Now, for the meat and potatoes of the book: diagnosing the vehicles. We will cover the Big Three, as well as Asian and European vehicles. You want faults? Well, we've got faults! All P0xxx type faults and an explanation are found in the Appendix A of this book. In addition, we offer diagnostic tips on some common problems with OBDII, and all the P1xxx type faults for GM, Ford and Daimler-Chrysler Vehicles. Read each chapter:

- **OBDII Diagnostic Tips for GM**
- **OBDII Diagnostic Tips for Ford**
- **OBDII Diagnostic Tips for Daimler Chrysler**.

OBDII in Europe? Yep, it is called **EOBD**. I think you can figure out what that means. Hint, E equals Europe… OBDII reached Europe to help clean up the air over there also. Don't think that the ozone layer is only affected in the US.

Where will OBDII go in the future? Find out in the chapter **Future of OBDII**. Will there ever be an OBDIII?

After a hard day of repairing cars, it is nice to come home to **Justina's Kitchen** and have a wonderful meal. (If your wife is upset that you bought yet another automotive book, just show her this chapter…).

Appendix A is a complete list of OBDII Generic fault codes. This is not just a simple list, it is also diagnostic help for each fault. With each fault, see a brief diagnostic tip to guide you in the right direction for diagnosis.

Appendix B is a complete list of Acronyms that OBDII adapted from the SAE.

Appendix C is a complete list of Terms and definitions that OBDII adapted from the SAE.

Ever wondered what makes an automotive engine run? In this chapter you will get an understanding of how an engine really works. If you are not familiar with the basics of an engine, OBDII will not make much sense. We are not going to get into complex issues, just simple facts.

Chapter 1 > Basic Concepts of an Engine

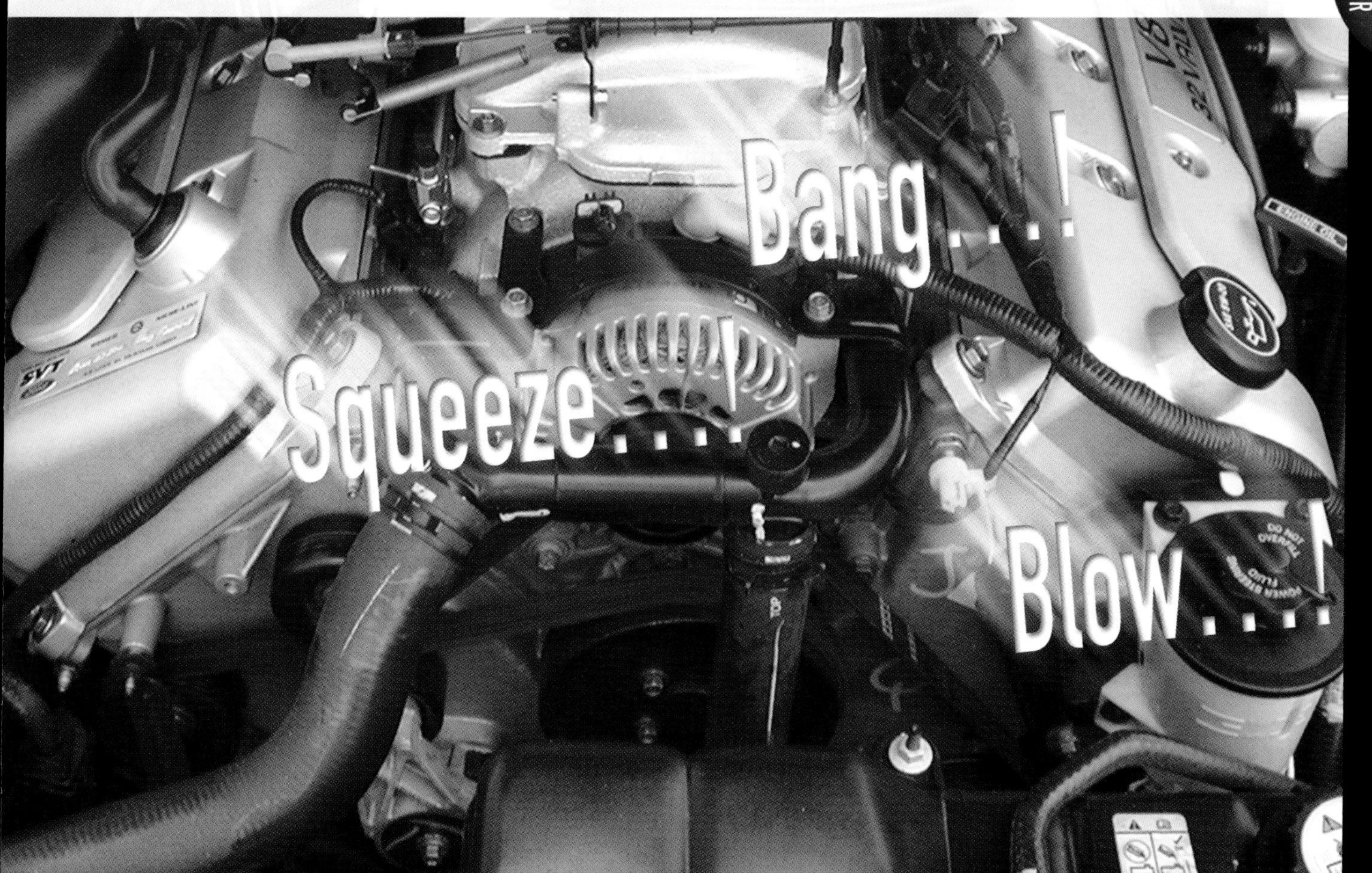

Suck, Squeeze, Bang, Blow!

First of all, get your mind out of the gutter. This phrase was used the first time I learned about engines. I was 13 years old, when my father (out of complete frustration with my taking apart the lawnmower, vacuum cleaner, etc.) took me to a lawnmower shop for a job that changed my life. The owner, John Anderson, taught me about the basics of engines.

IN LAYMAN'S TERMS, a lawnmower engine uses the same principles that a Ferrari engine uses: Suck, Squeeze, Bang, Blow. These are the principles of the four-cycles that a gasoline engine uses. If you are missing one of these, I don't care what vehicle you have, you are not going anywhere.

Suck or the Intake Stroke

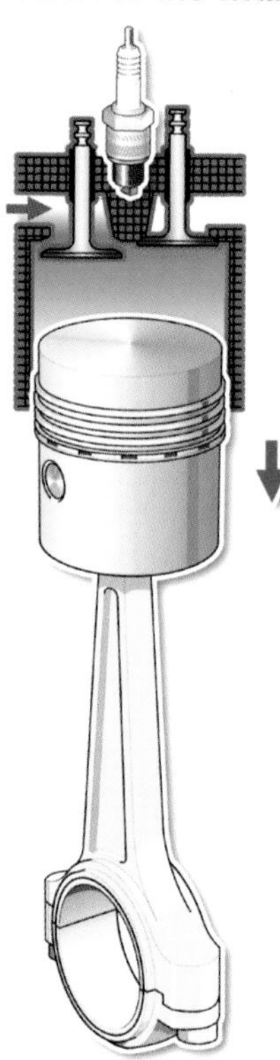

In order for an engine to work, it must have fuel. That is what the Intake Stroke does. When the ignition key turns in your car, the starter turns the crankshaft, which then makes the piston move. The piston is drawn down, like a plunger in a syringe. When this happens, the intake valve opens, and a proper mixture of fuel and air is then drawn into the cylinder. OBDII monitors that mixture. How the mixture is made, and carried into the cylinder, is the complicated part of an engine; you can have a carburetor or fuel injection. We will not go into all the details of how each works, since that could be a book in itself.

Squeeze or the Compression Stroke

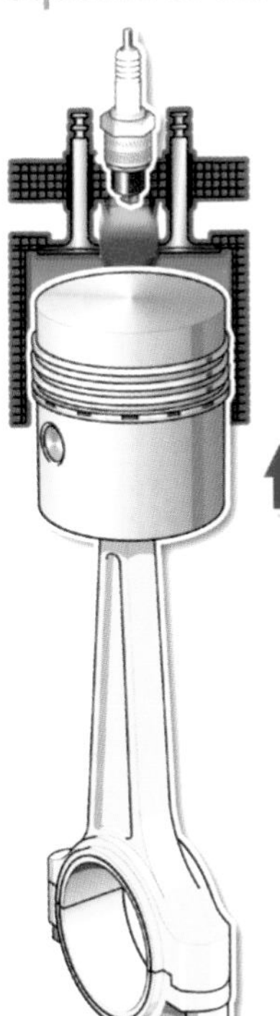

Now that the piston is drawn down, it will return back up, and at the same time, the Intake Valve will close. The cylinder will then build up pressure. If you take a syringe (PLEASE REMOVE THE NEEDLE BEFORE YOU TRY THIS), draw the plunger down, then put your finger over the opening at the top, and push the plunger up. Builds a lot of pressure, huh? The same will happen with an engine cylinder. This is important because this process will prepare the fuel and air mixture for the next stroke. In a diesel engine, this is most important since it depends on compression to combust.

Bang or the Combustion Stroke

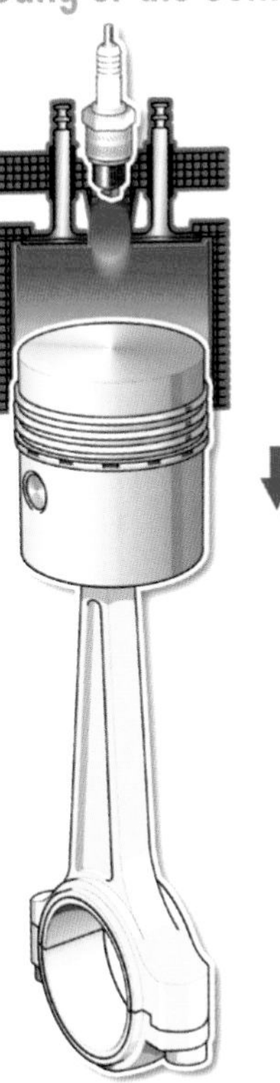

When the piston comes back to the top of its stroke, known as Top Dead Center (TDC), the spark plug is then ignited. That is when the BANG will happen. With the mixture of fuel and air compressed, and a spark of 20,000 to 80,000 volts, do you think that will produce a Bang? Yup, it sure does! At this point, the piston is forced down at such a rate that it will start to come back up. Also at this point, the engine is usually running on its own power.

Blow or Exhaust Stroke

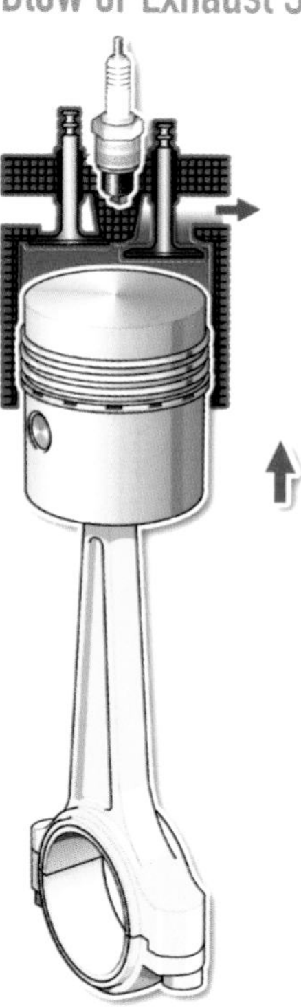

With the explosion that just happened, you will have burned fuel and air in the cylinder and you have to get rid of it. That is the key to OBDII, which monitors how much fuel and air is burned, and not burned. Please remember this later in the book, as we will refer back to this section. When the piston is on its way up (after being forced down), the Exhaust Valve will open, and all the burned fuel and air will then be forced out. When the piston comes back to the top, the Exhaust Valve will close, and the Intake Valve will start to open again. At this point, it is called Bottom Dead Center (or BTC). The entire process will then start over, and hopefully, for your sake, continue until you turn off the engine (or you will be walking).

Think about it - this process happens very fast and very often. Just look at the tachometer, which is measuring Revolutions Per Minute: if your engine is idling, it will indicate about 600 to 800 RPM. Therefore, the four strokes are happening 600 to 800 times per minute. At 60 MPH, usually the engine is in the area of 3000 RPM. That's 3000 explosions per minute! In just ten minutes there are more explosions than on July 4th in the entire USA!

Engine terminology used in this book

Before we start, we would like to give you an overview of the terminology that you will find in this book. The definitions are not from a textbook; they are our definitions of each term. If you don't know what the term means even after it is explained, don't worry - it is explained in detail later in the book.

AIR or Air Injection Reaction System

This is the engine's secondary air management system. Its purpose is to pump air into the exhaust and heat up the Catalytic Converter in order to speed up the process of burning off any unburned fuel in the exhaust. This usually happens when the engine is in its warm-up process.

Ambient Temperature Sensor

A sensor, which measures the temperature of the air outside of the engine environment.

Cam Position Sensor

This sensor tells the PCM where the camshaft is.

CAT or Catalytic Converter

The Catalytic converter is one of the oldest components used in controlling the emissions. The CAT is that thing that the mechanics of the 70's and 80's used to hollow out "in order for the engine to run better". Just kidding, not that we have ever seen that… It is really a part of the exhaust system of the engine. It is a ceramic honeycomb, which gets very hot (about 1500°F) in order to burn off any unburned fuel (Hydrocarbons or HC and many other gases) going through the exhaust. Its main purpose is to reduce the emissions.

Closed Loop

When all the components of the engine management system are at their normal operation conditions, the PCM will consult all its inputs (i.e. coming from the engine, exhaust, air temperature…) and automatically regulate the performance of the engine.

Continuous Monitor

A monitor that runs continuously during normal vehicle operation. This monitor looks at a set of components that could make the engine run out of its emission range.

Crankshaft Position Sensor

The crankshaft position sensor has many more tasks than to just tell the PCM where it is. This sensor is used to sense the speed of the engine. Also, it is used to determine if there has been a misfire in the engine. This is done by measuring the speed of each combustion stroke. The PCM knows how much thrust each stroke produces, and if one of these strokes uses less thrust, then it determines that there has been a misfire.

Diagnostic Executive

Software program found in the PCM used by Ford and GM. This is responsible for scheduling diagnostic tests, recording test results, storing Diagnostic Trouble Codes and controlling the Malfunction Indicator Lamp.

DLC or Data Link Connector

A OBDII connector that is predetermined by the specifications of the SAE. It is an outside link to the diagnostic scan tool.

ETC or Engine Temperature Sensor
Pretty much speaks for itself, huh?

EVAP or Evaporative Emissions System
A system that regulates and controls any fuel vapor that might escape into the atmosphere. You might think: "How is fuel vapor going to escape?" The gas tank is the biggest source. It is a big vapor leak waiting to happen. This system controls the vapor pressure expansion so the tank won't explode on hot days. It also looks for vapor leaks, like gas caps off or seal split and it checks to see that no fuel lines are leaking.

EGR or Exhaust Gas Recirculation Valve
Another one of the oldest component used to control emissions. Again: it is that thing that the mechanics of the 70's and 80's used to remove or fill with solder "so the engine would run better". Again: just kidding, we have never seen anyone to do that… An EGR valve is used to control a portion of the exhaust to be recycled back into the engine. This is used to return any unburned fuel back into the engine.

Freeze Frame
An information from the Diagnostic Trouble Code that is stored in the PCM. It can be used to help diagnose when, where, and how the fault happened. Freeze Frame is required by OBDII.

IAC or Idle Air Control Motor
This is a motor, or sometimes a valve, that controls how the engine idles.

IAT or Intake Air Temperature Sensor
Self-explanatory, we think…

Inches of Water
A measurement standard commonly used to measure small pressures inside the EVAP system. One PSI is equal to 27.68 inches of water.

Knock Sensor
A sensor that meters how each cylinder is firing. If one cylinder is firing too soon or too late, or even if it fires twice, a knock or ping is produced. The sensor sends a signal to the PCM, which will either adjust the timing on the ignition or the camshaft. Bad fuel, improper cooling or carbon deposits are just some of the symptoms that the Knock sensor will sense.

Live Data
Components of the engine from which the scan tool will display values in real time.

LTFT or Long Term Fuel Trim
The PCM is always learning how your vehicle is being driven. If you drive slowly, it stores that information; if you drive fast, it also stores that. In addition, the PCM will also compensate or adapt for a part of the engine that has become defective. This is called Adaptation or Fuel Trim. LTFT values represent the average of Short Term Fuel Trim fuel corrections.

MAP or Manifold Absolute Pressure Sensor
This senses the pressure inside the intake manifold. It senses the difference between the atmospheric pressure and the pressure, or vacuum, inside the intake manifold. It is also used to measure the load on an engine.

MIL or Malfunction Indicator Lamp
The MIL's purpose is to tell the driver of the vehicle there is a problem with the engine management system.

MAF or Mass Airflow Sensor
Tells the PCM the mass of air entering the engine.

Misfire
A miss in the ignition firing process of cylinder combustion. Please refer to Crankshaft sensor to see how this is determined.

Non-Continuous Monitor
A monitor that runs a specific route then stops. This happens once per trip. This process looks at a set of components that could make the engine run out of its emission range.

Open Loop
When each of the components of the engine management system are not in their normal operation condition, the PCM will use predetermined, or default, values to regulate the performance of the engine. For an example, if the O2 sensor is not in its normal operating condition, the PCM will use a fixed value to control the fuel.

O2 or Oxygen Sensor
The oxygen sensor is positioned in the exhaust pipe and can detect rich and lean mixtures. If the engine is running rich (too much fuel), there is a lack of Oxygen in the exhaust. Likewise, if the engine is running lean, there is an abundance of Oxygen in the exhaust. The mechanism in most sensors involves a chemical reaction that generates a voltage. The PCM looks at the voltage to determine if the mixture is rich or lean, and adjusts the amount of fuel entering the engine accordingly.

PCM or Power-train Control Module

You have seen PCM mentioned here several times already. The PCM is the most important part of the engine management system. It controls everything that is happening. It is commonly known as the Brain box, Computer, control unit, ECU, EEC...

PID or Parameter Identification

A list of components (i.e. Live Data) that a scan tool will display.

STFT or Short Term Fuel Trim

Like the LTFT, the STFT adapts in order to keep the engine at stoichiometry in the closed loop mode.

Stoichiometry

The PERFECT or IDEAL air/fuel ratio for combustion. This is what all PCMs dream of. For gasoline, the ratio is 14.7:1.

Task Manager

Software program found in the PCM used by Daimler Chrysler. This is responsible for scheduling diagnostic tests, recording test results, storing Diagnostic Trouble Codes, and controlling the Malfunction Indicator Lamp.

TPS or Throttle Body

Monitors the throttle valve position (which determines how much air goes into the engine) so the PCM can respond quickly to changes, increasing or decreasing the fuel rate as necessary.

Vapor Canister

A canister that filters the fuel vapor as it is vented into the atmosphere.

VSS or Vehicle Speed Sensor

Again, self-explanatory...

This chapter is an excerpt from the Ford Tuning Secrets Revealed book. It is an introduction to the layman's world of automotive electronics. It should satisfy your curiosity about what is inside the computer box controlling your entire engine and OBDII diagnostics.

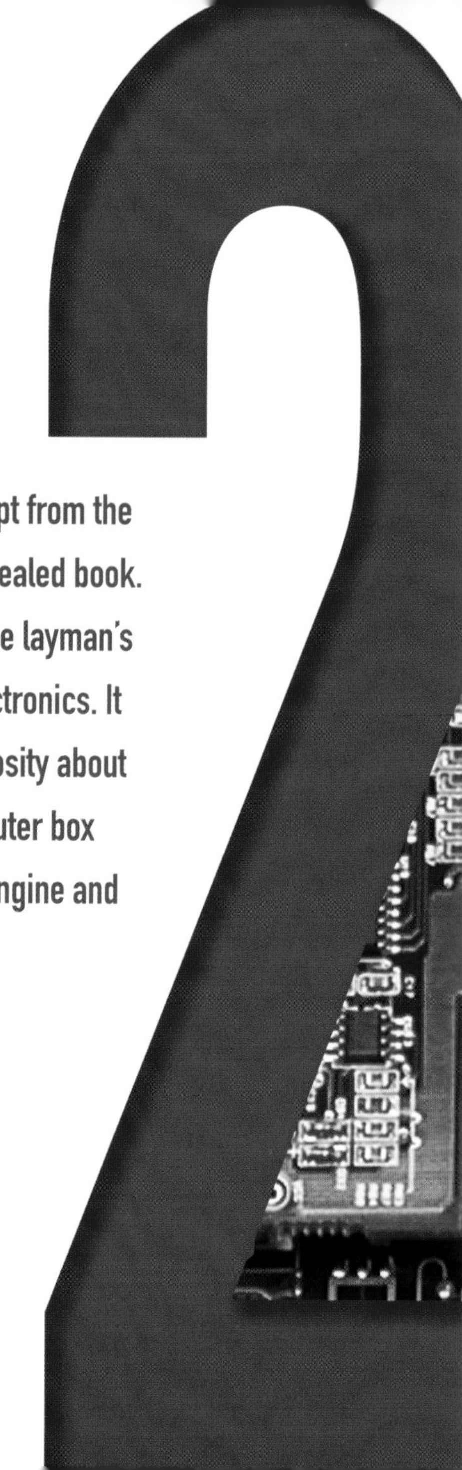

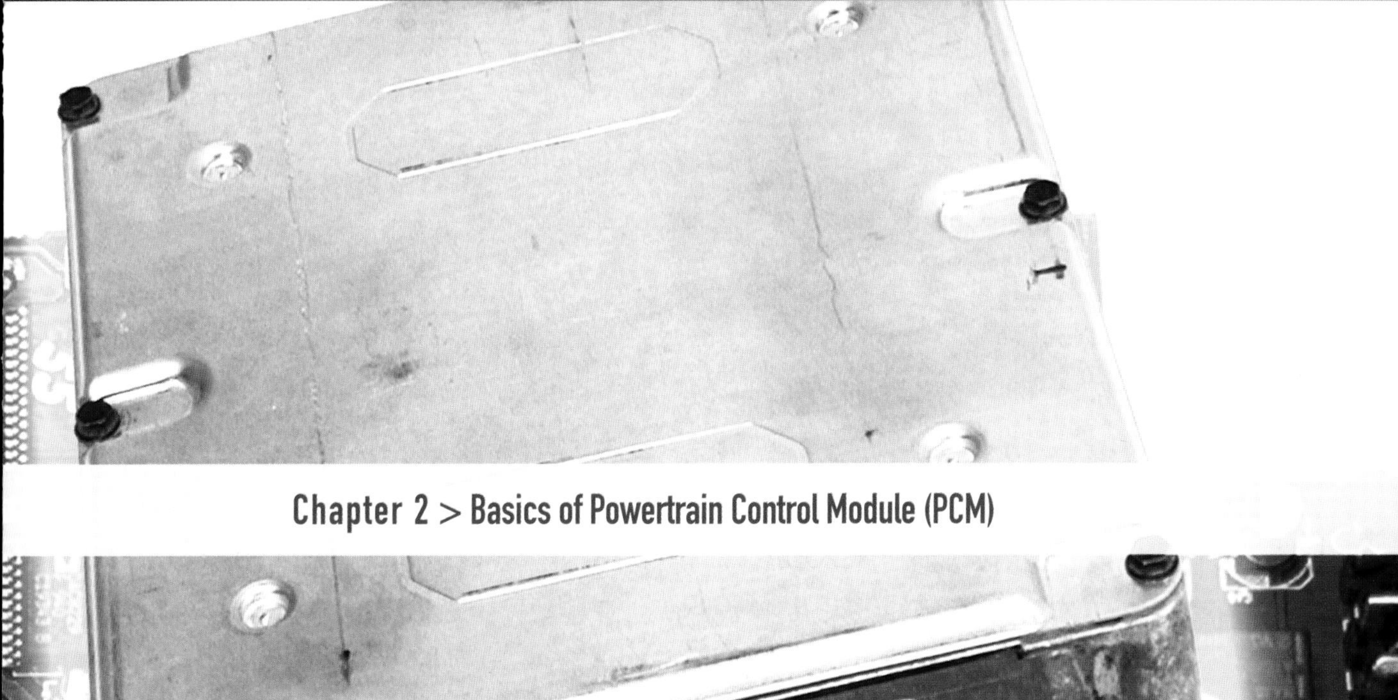

Chapter 2 > Basics of Powertrain Control Module (PCM)

Circuit Board and Components

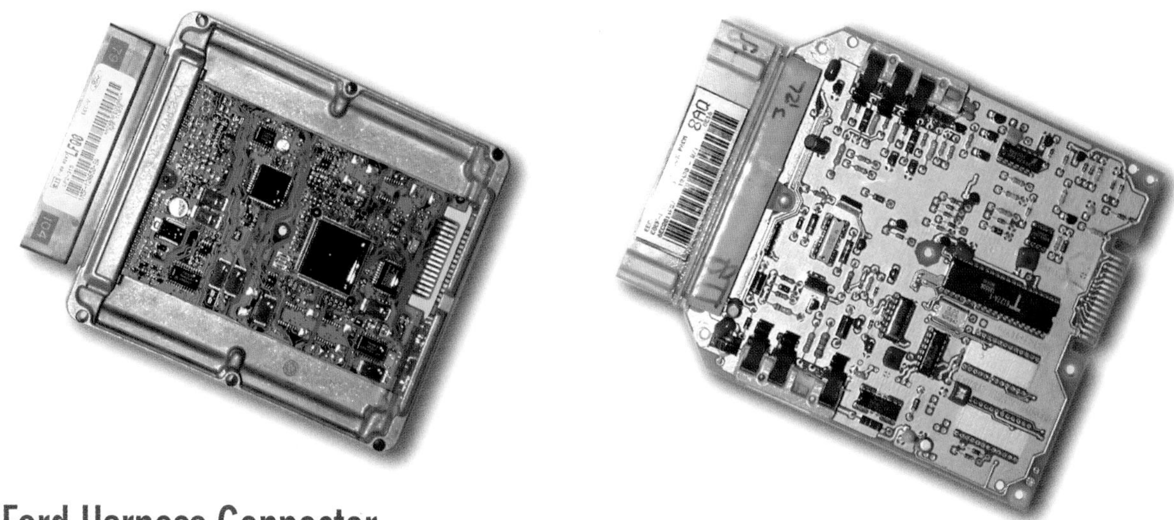

Ford Harness Connector

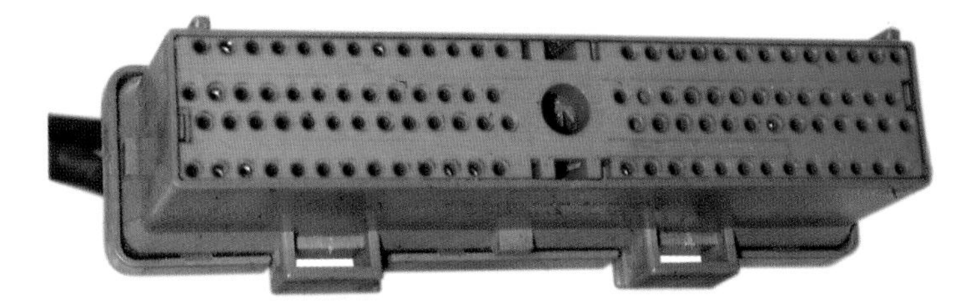

Microprocessor

"A *microprocessor* is a *Central Processing Unit* (CPU) packaged in an *Integrated Circuit* (IC, or better known as a "chip")."

Let's dissect the above definition.

What is a Central Processing Unit? It is the brain of any computer.

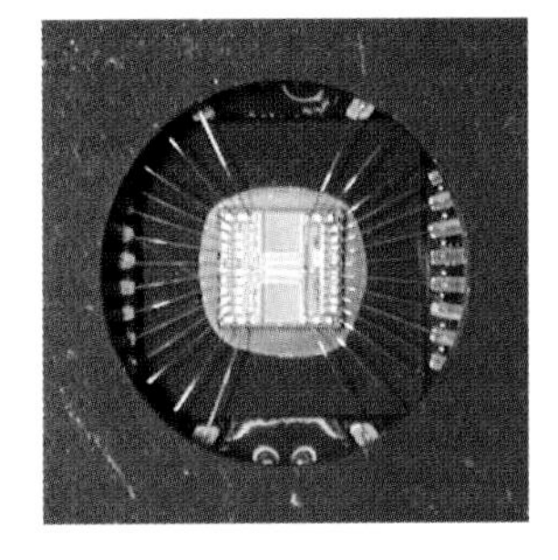

What is an Integrated Circuit? Commonly known as a computer chip, it looks like a black plastic bug with a bunch of legs soldered into a circuit board. In reality the chip is only a tiny silicon piece enclosed within the IC itself. This silicon chip contains tens of thousands of microscopic transistors. Transistors are basic building blocks of any electronic circuit.

The legs of the bug, which are called pins, are internally attached to the silicon chip. They protrude through the black plastic case in order to connect the silicon chip to the circuit board.

"The microprocessor executes instructions from a *program*, one at a time."

What is a program? Program is a procedure, written by humans, describing a task for a computer to execute. In our case, the task is to control the Ford engine.

What is a *human*? That is answered in some other publications.

IN LAYMAN'S TERMS, a microprocessor is an electronic brain of the PCM.

Read Only Memory (ROM)

"Read Only Memory (*ROM*) is an IC designed to permanently store the program and data for the microprocessor. This works like human long-term memory."

More descriptive name for this kind of a memory would be "One Time Write, then Read Only Memory", but that wouldn't sound very good, would it?

There are various types of memories falling in this category. Whether they are called ROM, *PROM* or *EPROM*, makes no difference to us, unless we are the computer designer.

Similarly to videotape, the information stored in ROM is *non-volatile*. That means it remains intact even if the power is off. Manufacturers of ROM claim a period of about 20 years of keeping data with the power off. Unlike the videotape, which wears down each time it is used, ROM gets refreshed for another 20 years each time it is powered up!

IN LAYMAN'S TERMS, the ROM contains the factory set engine control program. It includes the OBDII diagnostic procedures.

Random Access Memory (RAM)

"Random Access Memory (*RAM*) is an IC designed to temporarily store data used by the microprocessor. This works like human short-term memory."

A more descriptive name for this kind of memory would be the original name "Read and Write Memory". This name ended up as unused as the metric system in USA.

RAM will lose all data as soon as the ignition is turned off.

The PCM also contains *KRAM* (Keeper RAM or Keep-alive Memory or KAM). It is an ordinary RAM, powered directly by the car battery. KRAM keeps the data even if the ignition is turned off.

Both RAM and KRAM will lose data if the car battery dies or is removed.

IN LAYMAN'S TERMS, KRAM stores the OBDII faults.

Bus

Bus is another name for multiple signal lines dedicated to the same task. The bus is like an electrical cable. The purpose of the bus is to connect one chip to another. In real life, instead of wires, etched copper connections are used.

The count of lines within a bus is called *Bus Width*. Bus width is measured in *bits*. Typical bus is 8, 16 or 32-bits wide. This means that if a bus is 16-bit wide, it has 16 etched lines on the circuit board.

Bus connects the PCM's main chip (microprocessor), to other chips, like ROM and RAM, allowing them to share information.

The Integrated Circuit (IC, or simply a chip) sending the information through the bus is a transmitter. Engineers call it "chip writes to the bus". Likewise, the IC receiving the information from the bus is a receiver, or simply "reads the bus".

Similarly to the traffic on the road, which can be one-way or two-way; data can flow through the bus one-way or two-way. The bus is like the road and data is like the traffic. Engineers like to call the two-way bus "bi-directional".

This line is an etched copper connection.

These 8 lines are called a bus. The Bus Width pictured here equals 8.

A computer needs more than just one bus. To distinguish one from another, buses are named by the signal they carry. Two very common buses are *Address Bus* and a *Data Bus.*

IN LAYMAN'S TERMS, the bus is a bunch of wires linking one chip to another.

Purpose of the bus

Memory chips are able to store thousands of data pieces, called *bytes.* An example of a byte is one character. The English word "Hello" consist of 5 characters; therefore its size is 5 bytes.

Engineers call the whole content of the memory *data.* The individual pieces of the data are called *data bytes.* Each data byte is stored on a different location of the *memory chip.* These locations are called *addresses.*

In order to retrieve a byte from the memory, the microprocessor must first point to its location. The address bus is used to point to the location. The address bus is a one-way bus and only the microprocessor can write to it. One way to visualize it is that the address bus is a one-way road leading from the microprocessor.

The data bus is used to retrieve the data from the memory. The data travels from the memory to the microprocessor. Data bus can be used also to write new data to memory, in which case the data travels the other way - from the microprocessor to the memory. That means that the data bus is bi-directional bus.

IN LAYMAN'S TERMS, the purpose of the bus is to connect multiple chips to allow them to talk to each other.

In pre-OBD times there was no need for any bus. That was the Analog Age. One signal wire would carry a huge amount of combinations of data. Analog data changes amplitude over time. Measuring it with a voltmeter, it can have any value, in case of automobiles it is usually between 0 and 14.4 Volts (that is going to change very soon - your next vehicle will have a 36 Volt battery!). Example of an analog line is an audio output wire from the radio receiver to the speaker.

Today we live in the Digital Age. Digital data lines can unfortunately carry only one of two possible values. Measuring it with a voltmeter, in a 5 Volt digital system, it can have only value of 0 or 5 Volts. The actual values might differ slightly, but the digital circuit will ignore the variation and round it off to the nearest value. If the voltage is below 2.5 Volts, it will treat it as a logical 0, or else it will be treated as a value of logical 1. It is thanks to this ignorance that digital circuits are so resistant to noise. (Almost every electronic gadget around you operates digitally).

The PCM contains both analog and digital circuits. Most wires coming out of the PCM are connected to analog sensors (water temperature, mass-air meter, oxygen sensor, etc.) and analog electrical components (fuel pump, fuel injector, etc.).

3

The history of OBDII dates back all the way to 1955.

We will describe how, and why, all the Federal Acts and Mandates failed, and conclude how OBDII succeeded. The struggle of OBDII to come to maturity is quite eventful.

Chapter 3 > History of OBDII

1955

From the beginning until today

OBDII compatibility was incorporated into a few models already in 1994, including the Buick Regal 3800 V6; Corvette; Lexus ES3000; Toyota Camry (1MZ-FE 3.0L V6) and T100 pickup (3RZ-FE 2.7L four); Ford Thunderbird & Cougar 4.6L V8; and Mustang 3.8L V6.

1995 vehicles with OBDII include Chevy/GMC S, T-Series pickups, Blazer and Jimmy 4.3L V6; Ford Contour & Mercury Mystique 2.0L 4-cyl. & 2.6L V6; Chrysler Neon, Cirrus and Dodge Stratus; Eagle Talon 2.0L DOHC (non turbo); and Nissan Maxima and 240 SX.

Then, in 1996, federal regulations mandated OBDII to be used on all NEW cars and light trucks (domestic and imported) sold in the United States. By 1999, ALL vehicles had to have implemented the OBDII standards fully.

How did OBDII come about? In 1989, because of the continued failure of emission systems dating back to 1963, the Environmental Protection Agency (EPA, www.epa.gov) finally got together with the Air Resources Board (ARB, www.arb.ca.gov/homepage.htm) and perfected the practices that ARB had been doing for years. They wanted a better way to detect engine performance problems that caused emissions to rise. The goal was cleaner air and a better environment. This system would not replace emissions testing, but would act as an On-Board Diagnostic emissions monitor.

History, Year by Year

(for details, see the CARB web site www.arb.ca.gov/homepage.htm)

Most Pollution Regulations started in California, since it is the most populated and with the most cars. They are used as a model for the rest of the country.

1955

This is the year that started it all.

Federal Air Pollution Control Act of 1955 is enacted, providing for research and technical assistance and authorizing the Secretary of Health, Education and Welfare to work towards a better understanding of the causes and effects of air pollution.

Los Angeles County Motor Vehicle Pollution Control Laboratory begins within the Los Angeles APCD.

1959

California enacts legislation requiring the state Department of Public Health to establish air quality standards and necessary controls for motor vehicle emissions.

1960

California's population reaches 16 million people. Total registered vehicles approach 8 million and VMT is 71 billion.

The Motor Vehicle Pollution Control Board is established. Its primary function is to test and certify devices for installation on cars for sale in California.

Federal Motor Vehicle Act of 1960 is enacted requiring federal research to address pollution from motor vehicles.

1961

The first automotive emissions control technology in the nation, Positive Crankcase Ventilation (PCV), is mandated by the California Motor Vehicle State Bureau of Air Sanitation to control hydrocarbon crankcase emissions. PCV withdraws blow-by gases from the crankcase and re-burns them with the fresh air and fuel mixture in the cylinders.
This is the first of the mandated emission parts that are to be used in cars.

1963

PCV Requirement of 1961 goes into effect on domestic passenger vehicles for sale in

California. First Federal Clean Air Act of 1963 is enacted.

Empowers the Secretary of the federal Health, Education, and Welfare to define air quality criteria based on scientific studies. Provides grants to state and local air pollution control agencies.
In 1963 when OBDII was started, no one predicted that it would become what it is today.

1964
Chrysler exhaust control system is approved by the Motor Vehicle Pollution Control Board. Four other independent companies also received approvals.

1965
Federal Clean Air Act of 1963 is amended by the Motor Vehicle Air Pollution Control Act of 1965. Direct regulation of air pollution by the federal government is provided for, and the Department of Health, Education, and Welfare is directed to establish auto emission standards

1966
Auto tailpipe emission standards for hydrocarbons and carbon monoxide are adopted by the California Motor Vehicle Pollution Control Board. They are the first of their kind in the nation.

California Highway Patrol begins random roadside inspections of vehicle smog control devices.

1967
California Air Resources Board (CARB) is created from the merging of the California Motor Vehicle Pollution Control Board and the Bureau of Air Sanitation and its Laboratory. Enacting legislation is the Mulford-Carrell Air Resources Act, signed into law by Governor Ronald Reagan.

Federal Air Quality Act of 1967 is enacted. Establishes framework for defining "air quality control regions" based on meteorological and topographical factors of air pollution. Allows the State of California a waiver to set and enforce its own emissions standards for new vehicles based on California's unique need for more stringent controls.
The Birth of the CARB.

1970
California's population reaches 20 million people. Total registered vehicles exceed 12 million and VMT is 110 billion. Statewide average for NOX emissions per vehicle (new

and used) is 5.3 g/mile and 8.6 g/mile per vehicle for hydrocarbons. Cumulative California vehicle emissions for nitrogen oxides and hydrocarbons are about 1.6 million tons/year.

U.S. Environmental Protection Agency (U.S. EPA) created to protect all aspects of the environment.

Federal Clean Air Act Amendments of 1970 enacted. They serve as the principal source of statutory authority for controlling air pollution. This establishes basic U.S. program for controlling air pollution.
The year 1970 was a monumental and disappointing year for the environment. The EPA was established, and the Clean Air Act amendment started the ball rolling for OBDII. The problem was that no one wanted to obey all the rules. To do so would cost most companies millions of dollars; so most companies either ignored this or did exactly what the law stated; no more, no less. This created more problems, but the government didn't do much to enforce the law, and that created even more problems.

1975

First Two-Way Catalytic Converters come into use as part of Air Resources Board's (ARB) Motor Vehicle Emission Control Program.

California Air Pollution Control Officers Association (CAPCOA) created.

EPA Working Group established to develop strategies for State Implementation Plan activities.

1976

ARB limits lead in gasoline.

In 1977 Volvo introduces a car billed as "Smog-Free". Features the first Three-Way Catalytic Converter to control hydrocarbons, nitrogen oxides, and carbon monoxide.
Gasoline, as we know it, changed. A limit is put on leaded gasoline. Unleaded gas protects the Catalytic Converters from being destroyed, and helps the exhaust content from being emitted into the air. This was unfortunate for engines, but fortunate for our future.

1977

Federal Clean Air Act Amendments of 1977 enacted. This requires a review of all National Ambient Air Quality Standards by 1980.
This also was a pivotal year for the environment, since the EPA and ARB knew that the auto manufacturers were not doing what they were supposed to do. They are beginning to get strict.

1980

Compliance testing performed by ARB on autos in use to determine whether they continue to comply with emission standards as they age. This is a strong incentive for manufacturers to develop more durable emission control equipment to avoid the risk of recall.
The auto manufacturers are now under the gun. It is 1980, and not much has been done by the auto manufacturers. The study showed this and the following years would be harsh for them.

1982

ARB comes up with the first design for an On Board Diagnostic plan.
At last, OBD was born, but it still did not scare the auto manufacturers into developing the proper vehicle to pass emissions testing.

1988

California Clean Air Act is signed by Governor Deukmejian. Sets forth the framework for how air quality will be managed in California for the next 20 years.

ARB puts into effect On Board Diagnostics. Every vehicle sold in California must have this system.

ARB also adopts regulations effective on 1994 model cars requiring that they be equipped with on-board computer systems to monitor emission performance and alert owners when there is a problem.
OBDII was conceived and the auto manufacturers are not too happy, since they procrastinated for too long.

1990

ARB approves standards for Cleaner Burning Fuels and, Low and Zero Emission Vehicles.

The Clean Air Act Amendments of 1990 are signed into law by President George Bush. They rely largely on elements of the California Clean Air Act, and require a number of new programs aimed at curbing urban ozone, rural acid rain, stratospheric ozone, toxic air pollutant emissions and vehicle emissions, and establishes a new, uniform national permit system.

1992

Phase I of California Cleaner Burning Gasoline (CBG) comes to market. The result is 220 tons less ROG released every day (6% reduction), and elimination of the use of lead in gasoline. In November of the same year, ARB requires addition of oxygenates to gasoline to cut CO emissions by 10%.

1993

ARB enacts new standards for cleaner diesel fuel, resulting in a reduction of diesel particulate emissions by approximately 14 tons/day, 80 tons/day less sulfur dioxide and 70 tons/day nitrogen oxide emissions. Diesel buses and trucks are a major source of NOX. California Diesel Fuel comes to market.

1994

Smog Check II signed into law by Governor Wilson following lengthy negotiations with the federal EPA, designed to meet the requirements of the Federal Clean Air Act as amended in 1990. This program targets vehicles that pollute at least 2 to 25 times more than the average vehicle and requires repairs and re-testing of offending vehicles.

1996

The Big Three automakers commit to manufacture and sell Zero Emission Vehicles.

California Phase II Cleaner Burning Gasoline (CBG) comes to market. CBG reduces lung-damaging ozone and ozone precursors by 300 tons/day, as well as reducing airborne toxic chemicals like benzene that can cause cancer. This is equivalent to taking 3.5 million cars off the road.

California's State Implementation Plan (SIP) for ozone is approved by U.S. EPA on September 26, 1996.

1998

ARB identifies diesel particulate emissions as a Toxic Air Contaminant.

The ARB amended off-road engine regulations for lawn mowers, weed trimmers, and other small engine power tools.

The ARB adopted its LEVII emission standards for most mini vans, pickup trucks and sport utility vehicles (SUVs) up to 8,500 pounds gross vehicle weight to reduce emissions to passenger car levels by 2007.

Marine engine regulations were adopted to greatly reduce smog-forming emissions and water pollution from outboard engines and personal watercraft.

1999 - today

The California Fuel Cell Partnership, a public-private venture to demonstrate fuel cell vehicles in California, formally began. The Partnership includes auto manufacturers, energy providers, fuel cell manufacturers, and the State of California.

The ARB adopted a new regulation that reduces the smog-forming emissions from portable gas cans by over 70 percent.

Consumer products rules were adopted to cut smog-forming emissions and volatile organic compounds (VOC) from an estimated 2,500 common household products ranging from nail polish remover to glass cleaners.

The Board approved a new set of gasoline rules that will ban the additive MTBE while preserving all the air-quality benefits obtained from the state's cleaner-burning gasoline program.

Conclusion

Ok, now that you have all the dates, and their meanings, what does all this mean to you as an individual? Simple: the government found out in the late 50's that the automobiles they were supporting were essentially destroying our atmosphere. At first, they did not know what to do, so they kept coming up with committees to evaluate the situation. After years of committees and evaluations, laws began to come into effect. As these laws, or Acts, came into play, auto manufacturers ignored them, and put on items that the law required just to pass the test and the standards. When the 1970's hit, the lawmakers took a stand and again were ignored. 1977 was the first year that the lawmakers really made a difference. With the gas shortage, and other problems like recession and economic turmoil, the auto manufacturers had to do something to sell cars. The EPA and ARB were getting serious.

As you can see, the concept of OBDII has been around for years. In the past, each manufacturer used their own systems and controls. To resolve this, the Society of Automotive Engineers (SAE, www.sae.org) proposed several standards. The birth of the OBD was marked when the ARB mandated many of the SAE's standards for emissions in California on 1988 and later vehicles.

The original OBD system was not complex at all. It monitored the Oxygen sensor, Exhaust Gas Recirculation (EGR) system, fuel delivery system and the Engine Control Module (ECM) for excessive emissions. It did not require any uniformity from the manufacturers. Each vehicle manufacturer had their own procedure for monitoring emissions, as well as diagnosing the system when the emissions were out of the normal range. The emission monitoring systems were not working efficiently, since they were designed as a patch for vehicles already in production. Vehicles not originally designed for the emission systems failed drastically and the manufacturers were not necessarily complying with regulations. They did (more or less) what ARB and EPA told them to, but nothing more. Imagine if you were an independent repair facility. You had to have a unique diagnostic tool, manuals for codes, and repair manuals for each manufacturer. Vehicles were not being repaired properly, if at all.

The United States Government was besieged on all sides, from independent repair facilities to lobbyists for clean air. EPA was asked to step in. They took the standards of the SAE and the ideas of ARB and created an extensive list of procedures and standards. By 1996, all manufacturers had to comply in order to sell vehicles in the United States.
The second generation of On-Board Diagnostics was birthed, with the name of OBDII.

In the Anatomy of OBDII we will

dissect all the parts.

We start with the Diagnostic

Connector, Protocols,

Trouble Codes,

and finish

with Terminology.

Chapter 4 > The Anatomy of OBDII

ENTER

Basics

As you just read in the History chapter, OBDII is not some basic concept that was just thrown together. It took years of development. Please remember, OBDII is not an Engine Management System. It is a set of rules and regulations that each manufacturer must follow in order to have their Engine Management System pass Federal Emissions. To best understand what OBDII is we must break it down into sections. When doctors are studying, they don't just study the body as a whole, they study part by part. In the end, all the parts come together. That is what we are doing here. OBDII must have all of the following parts to make up the standardization.

One centralized diagnostic connector with specific pins with assigned specific functions

The diagnostic connector (OBDII calls it the Diagnostic Link Connector - DLC) main function is to enable the diagnostic scan tool to communicate with OBDII compliant control units. The DLC must follow the standards set by SAE J1962. According to J1962, the DLC must have a centralized location in the vehicle. It must be within 16 inches of the steering wheel. The manufacturer can place the DLC in one of eight possible places predetermined by the EPA.

The standard OBDII connector

Pin 1 - Proprietary
Pin 2 - J1850 Bus+
Pin 3 - Proprietary
Pin 4 - Chassis Ground
Pin 5 - Signal Ground
Pin 6 - CAN High (J-2284)
Pin 7 - ISO 9141-2 (K Line)
Pin 8 - Proprietary

Pin 9 - Proprietary
Pin 10 - J1850 Bus-
Pin 11 - Proprietary
Pin 12 - Proprietary
Pin 13 - Proprietary
Pin 14 - CAN Low (J-2284)
Pin 15 - ISO 9141-2 (L Line)
Pin 16 - Battery Power

Each pin has its assigned definition. Assignment of many of the pins is still left up to the manufacturer, but those pins are not meant to be used by OBDII compliant control units. They can be for Supplemental Restraint Systems (SRS), or Anti-Lock Brake Systems (ABS) to name just two.

You may have noticed the connector has its own power and ground source (pins 4, 5 are ground, pin 16 is power). This is so that a scan tool won't require an external power source. If you plug in a diagnostic scan tool, and it doesn't power up, first check pin 16 for power, and then pins 4 and 5 for a ground.

You probably noticed also these alphanumeric characters: J1850, CAN and ISO 9141-2. These are protocol standards developed by SAE and ISO. The manufacturers have their choice of these standards to use for their diagnostic communication.

Each standard has a specific pin to communicate on. For example, Ford products communicate on pins 2, and 10. GM products communicate on pin 2. Most Asian and European products communicate on pin 7, and some, also, on pin 15

Which protocol is used makes no difference to understanding OBDII. The message exchanged between the diagnostic tool and the control unit is always exactly the same, only the way it is transmitted differs.

IN LAYMAN'S TERMS, one standardized connector, with one shape, in one location, makes it easier and cheaper for the repair shops. They don't need 20 different connectors or tools for 20 different vehicles. In addition, this saves time, since the repair shops won't have to hunt down the location of the connector to hook up the tool.

Standardized diagnostic communication protocols

As seen above, OBDII recognizes several different protocols. At this point, we only need to discuss three of them: J1850-VPW; J1850-PWM; and ISO9141, which directly affect all the vehicles in the United States.

All the control units in the vehicle are connected with a cable (called a diagnostic bus), creating a network. We can connect a Diagnostic Scan Tool to the diagnostic bus. The tool will send out a signal to the specific control unit to which it wants to communicate. The control unit will respond. Communication will continue until the tool terminates communication or the tool is disconnected.

For instance, the Diagnostic Scan Tool will ask the control unit "What are your faults?" The control unit will answer appropriately. With that simple communication exchange, we have just followed a protocol.

IN LAYMAN'S TERMS, protocol is a set of rules that must be followed in order for a network to complete a communication.

Classification of a Protocol

SAE has defined three distinct protocol Classifications, Class A, Class B, and Class C.

Class A is the slowest of the three and can be as high as 10,000 bytes per second or 10Kb/s. The ISO 9141 standard uses the Class A protocol.

Class B is ten times faster and supports communication of data as high as 100Kb/s. The SAE J1850 Standard is a Class B protocol.

Class C supports communication performance as high as 1Mb/s. The most widely used vehicle-networking standard for Class C is Controller Area Network (CAN). Higher performance communication classifications from 1Mb/s to 10Mb/s are expected in the future. Classifications like Class D can be expected as bandwidth and performance needs go forward. With Class C, and the futuristic Class D protocols, we will be able to use fiber optics as cabling for the network.

J1850 PWM protocol

J1850 comes in two different flavors. The first is a high speed 41.6 Kb/s Pulse Width Modulation (PWM). Ford, Jaguar and Mazda use this. Ford was the first to use this type of communication. The communication uses two wires, pins 2 and 10 of the diagnostic connector.

J1850 VPW protocol

The other J1850 alternative is the 10.4 Kb/s Variable Pulse Width (VPW). Both General Motors (GM) and Chrysler use this protocol. It is very similar to Ford's protocol, but the communication is much slower. It uses one wire, pin 2 of the diagnostic connector.

ISO 9141 protocol

The third protocol is the ISO 9141, defined by the International Standard Organization (ISO). Most European, Asian, and some Chrysler vehicles use this standard.

It is not as complex as the J1850 standards. While the J1850 protocols require use of specialized communication microprocessors, ISO 9141 uses standard off-the-shelf serial communication chips.

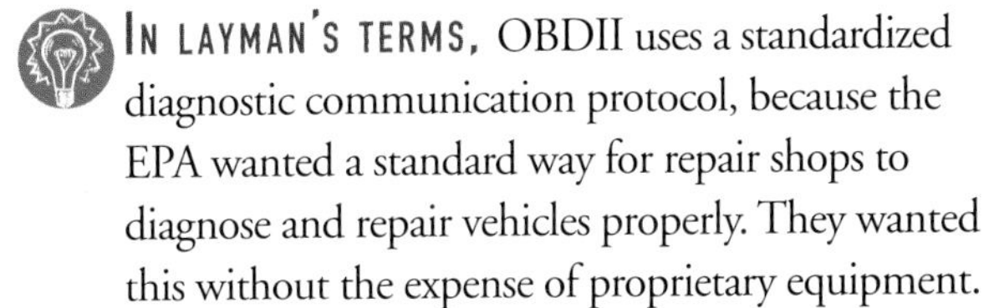

IN LAYMAN'S TERMS, OBDII uses a standardized diagnostic communication protocol, because the EPA wanted a standard way for repair shops to diagnose and repair vehicles properly. They wanted this without the expense of proprietary equipment.

For more intense description of the protocols, please refer to Chapter 4, Understanding OBDII Protocol.

Malfunction Indication Lamp (MIL) Operation

When the engine management has detected an emission related problem, the Check Engine light will come on. OBDII calls the light a Malfunction Indicator Light (MIL). The MIL will typically display the phrase "Service Engine Soon," "Check Engine" or "Check."

The MIL's purpose is to tell the driver of the vehicle there is a problem with the engine management system. If the MIL comes on, please don't panic. It is not life threatening. Your engine will not blow up. If the oil light or the overheat warning light comes on, then it is time to panic. The OBDII MIL is just telling the driver that there is a problem with the engine management system that will cause excessive emissions from the tailpipe or the evaporative fuel systems.

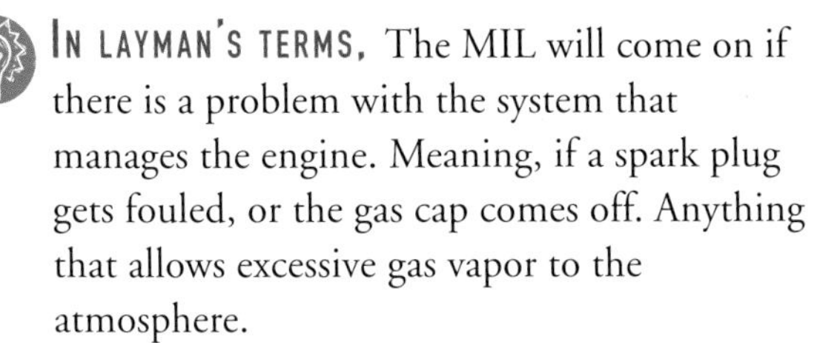

In layman's terms, The MIL will come on if there is a problem with the system that manages the engine. Meaning, if a spark plug gets fouled, or the gas cap comes off. Anything that allows excessive gas vapor to the atmosphere.

To check the function of the MIL: Turn the ignition to the run position (when all the lights come on in the dashboard), the MIL will illuminate. According to OBDII the light must come on for a period of time. Some manufacturers will have the MIL stay on, and others will come on then after a period of time it will turn off. If you start the engine, and all the conditions are met and no faults are found, the MIL will turn off.

The MIL will not necessarily come on when a fault first happens. The importance of the fault will determine when the MIL will come on. If the importance is listed as high, meaning if you have a gross petulant, then the MIL will come on immediately. The fault is listed as Active. Or, if the fault is a problem, but not enough to exceed the gross pollutant status, then the light will not come on, but will be placed in Stored status. In order for the fault to be listed as Active, it must happen on several occasions, called drive cycles. Typically a drive cycle is when a vehicle is started cold and driven to normal operating temperature (with coolant temperature below 122 F and the coolant and air temperature sensors within 11 degrees of one another). During this process all the on-board emission monitor tests must be completed.

Different vehicles have different engine sizes, and the drive cycle is slightly different for each. For further description of a drive cycle, please refer to the GM, Ford and Chrysler diagnostic tips section. On an average vehicle, if a fault is seen in 3 drive cycles then the MIL will come on. On the other hand, if the fault is not seen in 3 drive cycles, then the MIL will go out.

So, if the MIL comes on and goes out, don't worry. The fault remains stored in the PCM, and can be retrieved with a scan tool. As mentioned, there are two states that the fault can be in; Stored or Active. Stored is when the fault has been detected, but the MIL is not on, or was on and went out. Active is when the fault is present, and the MIL is on.

A Standardized set of Diagnostic Trouble Codes (DTC)

OBDII calls a fault a Diagnostic Trouble Code (DTC). A DTC is made up of a combination of 1 letter and 4 digits, as designated by SAE J2012. The figure below shows what each character means.

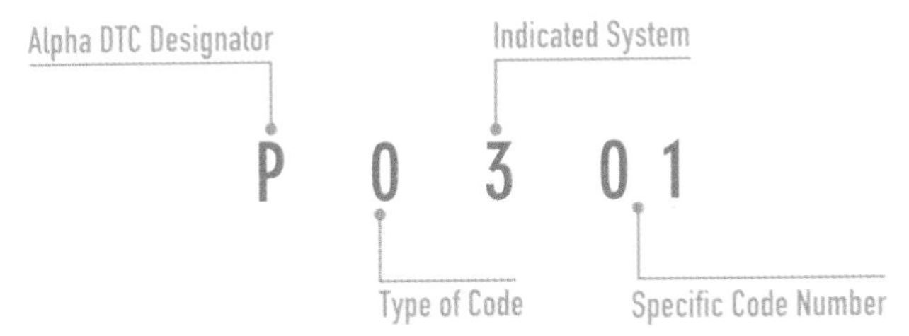

DTC Alpha Designator

As you can see, each character has a meaning. The first character is the DTC Alpha Designator. This letter identifies in which part of the vehicle the fault is found. The assignment of the letter (P, B, C, or U) is determined by the control unit being diagnosed. If two control units respond, the one with priority is listed. There can only be four letters found in this position:

P = Powertrain: these faults are related to the Engine, and Transmission

B = Body: faults that are related to

the Body of the Vehicle (i.e. SRS, Instrument cluster)

C = Chassis: faults that are related to the Chassis of the Vehicle (i.e. ABS, Traction systems)

U = Network Communication: faults that are related to the CAN

Type of Code

The second character is the most controversial - it shows who defined the code:

Value 0 (commonly known as "P0" code) means it is a generic, publicly known SAE fault code (see Appendix A for a complete list of generic P0 fault codes).

Value 1 (commonly known as "P1" code) means it is a proprietary fault code, defined by the vehicle manufacturer (see appropriate chapters for a complete list of P1 fault codes).

Most scanners are unable to recognize the description or text of the P1 codes. However, Hellion by DiabloSport (www.eobd.com) is one of the few that can recognize most of them. SAE made up the original DTC list, but the manufacturers complained that they had their own systems and each system was different. Mercedes has a completely different system than Honda, etc. They cannot use each other's codes, so the SAE made a provision to separate the standard codes (P0 codes) from manufacturer specific codes (P1 codes).

Indicated System

The third character is the Indicated System. This character is one of the least known, and most helpful. It will tell you the affected system, without even looking up the fault text. It helps to quickly identify the general area of the problem without knowing the exact description of the code. The numbers that are found here are as follows, if the number is:

1 = Faults that relate to the Air/Fuel Control Systems
2 = Faults related to the Fuel System (i.e. injectors)
3 = Faults that deal with Ignition System/ Misfire
4 = Faults related to the Auxiliary Emission Controls (i.e. EGR, AIR, CAT, EVAP)
5 = Faults related to the Vehicle Speed/ Idle Control and Auxiliary Inputs
6 = Faults related to the Computer Systems (i.e. PCM or the CAN)

7 = Transmission or Transaxle
8 = Transmission or Transaxle

Specific Code Number

The fourth and fifth characters are usually viewed together. These usually relate to the older OBDI fault code. OBDI DTCs are usually two digit faults. OBDII just took the two digit codes and put them on the end of the fault. This made it easier to distinguish the faults.

Now that we know what makes up the DTC, lets take the DTC P0301 as an example. Without even looking up the text, we have an idea of what the fault is. First, by the "P", we know it is an engine fault. Next is the "0", we know that it is a generic fault. Then we have a 3, by looking at the above paragraph, we will see that the "3" relates to the Ignition System/ Misfire. Last, we have the "01". In this case, the last two numbers tell you which cylinder is misfiring. Put this together, and we have an engine fault, with cylinder number three misfiring. If it were P0300, then it is a multiple misfire; meaning there are too many cylinders misfiring, and the control unit can't determine which one it is.

Specific self-diagnostic on-board monitoring of emission malfunctions

Wow, that sounds complex. It really is not! Ford and GM call this the Diagnostic Executive. Daimler Chrysler calls it the Task Manager. This is OBDII compliant software, which is a program inside the PCM that observes all that is happening. The PCM is a workhorse. It performs immense calculations in microseconds. It has to control: when the injectors open and close, when to fire the coil, when to advance the timing on the cams, and in the ignition and so on. While all this is happening, the OBDII software is checking on all these components to see if they are operating properly. The software:

- Controls the status of the MIL
- Stores DTCs and the Freeze Frame dat
- Checks the drive cycles that control the DTCs
- Starts and runs the monitors for the components
- Prioritizes the monitors
- Updates the readiness status for all the monitors
- Displays the tests results for the monitors

• Makes sure that the monitors don't conflict with each other

As seen above, in order for the software to complete its job, it must perform and complete monitors on the engine management system. What is a monitor? Simply put, a monitor is a test run by the OBDII software in the PCM to check on emission related components to see if they are running properly. According to OBDII, there are two types of monitors:

• Continuous Monitor - this monitor continuously runs once its conditions are met.
• Non-Continuous Monitor - this is a monitor that is run only once per trip.

For a complete list and explanation of all monitors that are being done, please refer to chapter OBDII Scan Tools.

> Monitors are important to OBDII. Like heart monitors, OBDII monitors are designed to test specific components in order to catch a malfunction or deterioration of the component that is being monitored. When a test fails, a DTC is then put into the PCMs memory.

Standardized component

As in any trade, there are different names and slang for the same word. Take fault code for instance. Some call it a code, others call it a fault, and still others call it "that thang that's broke". A DTC is a fault, code, or that " thang that's broke".

Before OBDII, each manufacturer had their own name for each component. I failed the first SAE test I took, since I was familiar with the terms used with European vehicles. The SAE tests used terms used with American vehicles. Now, since OBDII, all vehicles must use the same name for each component, and life is much easier for repairing cars and ordering parts.

Emissions terminology

As always when a governmental organization is involved, acronyms, and jargon is a must. The SAE came up with a standardized list of terminology for the components that pertain to OBDII. The standard is J1930. A complete list of Acronyms are found in Appendix B, and a list of Terms and their definitions are in Appendix C.

When you hear some people talk about protocols and wavelengths, you tend to get sleepy, but not in this chapter! You will be enlightened by learning about the pulse widths, modulations and high-speed communications.

This chapter is a basic overview of all four possible bus interfaces used in OBDII compatible vehicles.

Chapter 5 > Understanding OBDII Protocols

contributed by Ivan J. Kotzig

ENTER

Introduction

In the following pages, two terms (the protocol and the bus) are used seemingly interchangeably. This always confuses all the beginners studying OBDII.

"Please tell me what is J1850 - is it a bus or a protocol?" is a common question. The proper answer is that it can be both; yet, there is a big difference!

IN LAYMAN'S TERMS, when we talk about the **bus** or bus interface, we mean all the electronic circuits (hardware) involved in connecting a diagnostic scanner to a vehicle. When we talk about the **protocols**, we mean all the information (software) traveling between the scanner and the vehicle.

Let's overview all four possible bus interfaces and protocols used in the harsh automotive environment of OBDII compatible vehicles.

ISO9141

The ISO9141 automotive bus and protocol are mostly found in European automobiles. ISO9141 operates at data rates of 10.4 Kilobits per second.

It is the simplest one to implement of all four possible OBDII interfaces, because it is based on a standard serial interface found in all PCs. The PC serial connector is (in recent years) always a DB9 male connector, and the interface is called RS232.

ISO9141 uses two wires to communicate: K and L. K is a bi-directional signal, meaning that data can travel between the scanner and the vehicle in both directions. L is a single directional signal, meaning that data can travel only from the scanner to the vehicle.

If one would monitor the ISO9141 communication and connect an oscilloscope to K wire, the following signal waveform would appear:

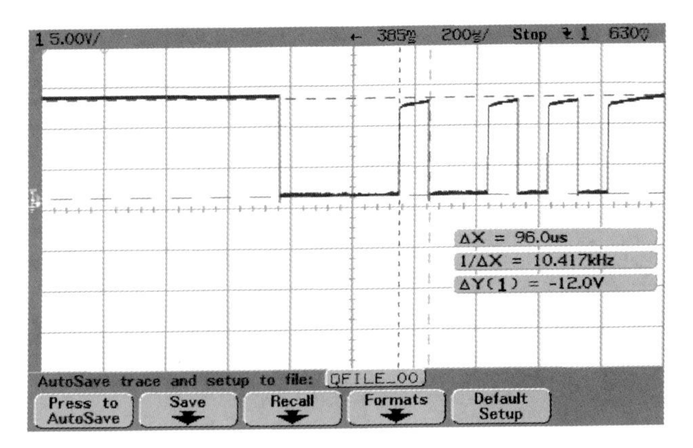

The ISO9141 bus cannot be directly connected to a PC, since the Voltage levels are different:

- The RS232 data from a PC travels as a square wave that can have two different voltage levels, +10V and -10V.

- The ISO9141 data travels as a square wave that can have two different voltage levels, +12V and 0V (ground).

This can be solved by a very simple voltage level shifter circuit - a common thing in electronics.

IN LAYMAN'S TERMS, once the voltage levels are converted, there is no other hardware necessary to connect an ISO9141 vehicle to a PC, and that makes it very easy to build a diagnostic tool! Although not trivial, all is needed is the software.

J1850 VPW

The VPW name stands for Variable Pulse Width. It is most commonly, but not exclusively, used on vehicles manufactured by GM. J1850 VPW operates at data rates of 10.4 Kilobits per second.

J1850 VPW uses only a single wire to communicate. It is a bi-directional signal, meaning that data can travel between the scanner and the vehicle in both directions.

If one would monitor the VPW communication and connect an oscilloscope to both bus wires, the following signal waveform would appear:

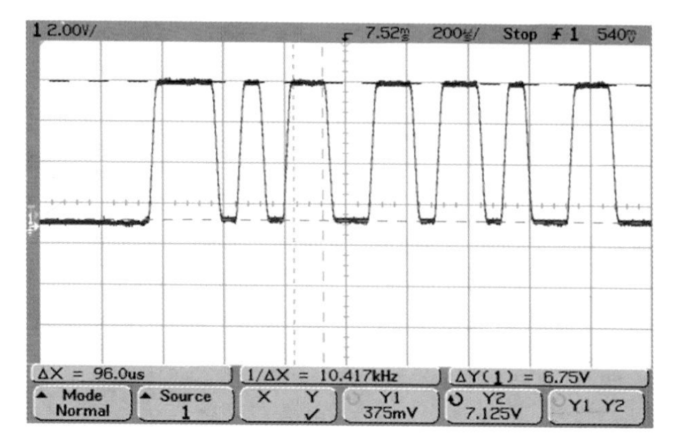

The J1850 VPW bus is unfortunately not based on the common RS232 based serial communication, which makes it much more difficult to interface the vehicle to a common PC. An intelligent interface is needed.

An intelligent interface is not just a simple voltage level shifter, as the one used in ISO9141. It has to be able to convert entire data messages between two very different serial communication concepts.

IN LAYMAN'S TERMS, one could compare the intelligent interface to a language translator. Let's say, we want to translate a conversation between Spanish and English businessmen. The Spanish buyer will ask a question "How much does it cost?" The translator must wait until the question is over, than translate it in his/her mind and, only then, pronounce it in English. The same is true in the other direction, when the American salesman says: "It all depends..."

Similarly, when the PC sends a request to the vehicle "Give me the current RPM", the interface receives it first, translates it into another format (in this case to J1850 VPW) and only then transmits it to the vehicle. The same is true for the other direction.

There are several concepts available for building an RS232 to J1850 VPW interface.

One popular solution is to use a dedicated integrated circuit manufactured by Motorola - MC68HC58.

MC68HC58 converts the OBDII bus into a microprocessor compatible bus, which is still not compatible with the RS232 found on ordinary PCs. There is one more conversion necessary, which is solved by using one of the many available microprocessors.

Unfortunately, that will create another problem: microprocessors require software. Software, in turn, requires a human programmer to write the program and that is a real pain in the wallet.

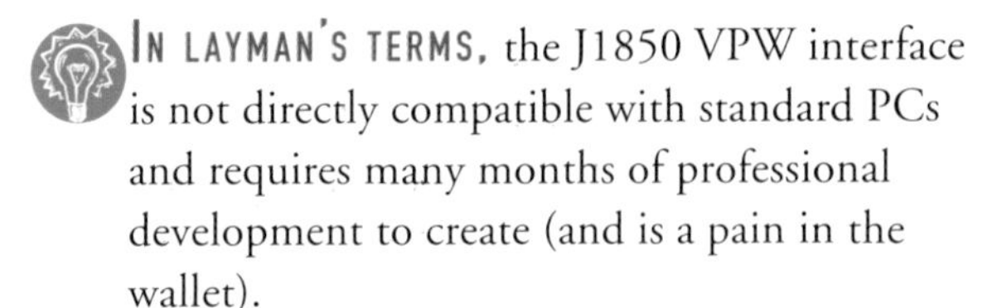

IN LAYMAN'S TERMS, the J1850 VPW interface is not directly compatible with standard PCs and requires many months of professional development to create (and is a pain in the wallet).

J1850 PWM

The PWM name stands for Pulse Width Modulation. It is most commonly, but not exclusively, used on Ford vehicles. J1850 PWM operates at data rates of 41.6 Kilobits per second.

The PWM uses two wires to communicate: Bus+ and Bus-. Both wires carry a bi-directional signal, meaning that data can travel between the scanner and the vehicle in both directions.

If one would monitor the PWM communication and connect an oscilloscope to both bus wires, the following signal waveforms would appear:

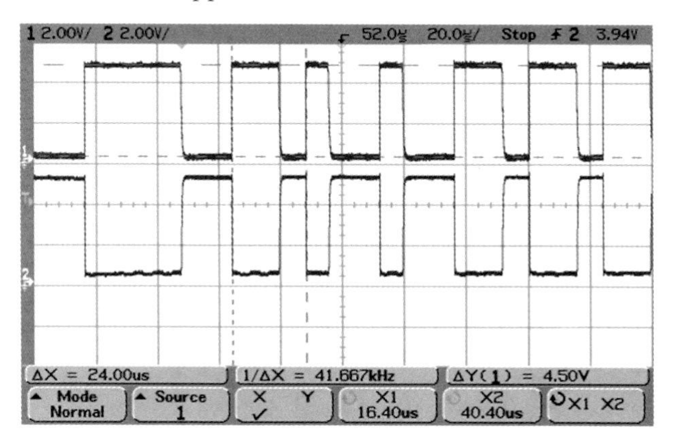

PWM is similar to a Morse code - the data is encoded with a series of predefined long and short pulses. Notice that one signal is exact inversion of the other. Engineers call these signals differential, and that gave the name PWM the differential bus.

One could argue why have two wires for the PWM bus, if they both carry practically the same information? Why not save and use just one wire as the VPW bus does. That is a good one! However, engineers know what they are doing and the differential bus is very resistant to electrical noise from the outside world.

Imagine the vehicle is driven near a radio station. The air is filled with electromagnetic noise, which could cause fake pulses in all wires in the vehicle and interfere with the communication. In a one-wire interface, that could easily confuse the electronic system in the vehicle, which could in turn interpret it as a command coming from the diagnostic scanner.

Differential bus helps to eliminate the confusion caused by electromagnetic interference. Electromagnetic noise creates the SAME fake pulses in EVERY wire, but the system accepts only pulses of opposite values. The bus will not see the noise and will not get confused. Isn't that cool?

In layman's terms, single wire bus is analogous to AM radio, and the differential bus is analogous to FM radio. While listening to music on AM, any small electromagnetic noise is added to the sound and can be heard through the speakers. FM radio waves are immune to the noise and the speakers rock (and roll)!

J1850 PWM bus is not based on the common RS232 based serial communication, which makes it also much more difficult to interface a PC to the vehicle. An intelligent interface, similar to the previously described VPW interface, is needed.

There are several concepts available for building an RS232 to J1850 PWM interface. One popular solution is to use a dedicated QBIC integrated circuit manufactured by Motorola. QBIC chip removes all the headaches from the design of otherwise complicated PWM interface.

The same problem already described in the section about VPW is true here as well: QBIC converts the OBDII bus into a microprocessor compatible bus, which is still not compatible with the RS232 found on ordinary PCs. There is one more conversion necessary, which is solved by using one of the many available microprocessors.

Unfortunately, microprocessors require software. Software, in turn, requires a human programmer to write the program and that is a real pain in the wallet.

In layman's terms, the J1850 VPW interface is not directly compatible with standard PCs and requires many months of professional development to create (and is a pain in the wallet). This makes us to like the good old ISO9141 even more!

CAN

The CAN name stands for Controller Area Network. It is the newest kid on the OBDII block. Originally developed by Bosch, it is now a recognized international standard ISO11898.

CAN overcomes the main problem of the other three OBDII protocols - the slow speed. CAN operates at data rates of up to 1 Megabits per second, which is about 100 times faster than ISO9141!

The CAN uses two wires to communicate: CAN+ and CAN-. Both wires carry a bi-directional signal, meaning that data can travel between the scanner and the vehicle in both directions. CAN bus is also a differential bus, which is resistant to electromagnetic interference (as explained in the section about J1850 PWM).

CAN bus is not based on the common RS232 based serial communication, which makes it also much more difficult to interface a PC to the vehicle. An intelligent interface similar to the previously described VPW and PWM interface is needed.

There are several chips available for building an RS232 to CAN interface:

- Cygnal (http://www.cygnal.com/products/productguide.asp)
- Fujitsu (http://www.fujitsu-fme.com/index4.html?/products/micro/can/start.html)
- Infineon (http://www.infineon.com/cgi/ecrm.dll/ecrm/scripts/prod_cat.jsp?oid=-8141)
- Intel (http://developer.intel.com/design/auto/can/prodbref/27270401.HTM)
- Philips (http://www.semiconductors.philips.com/buses/can/)
- Microchip (http://www.microchip.com/1000/pline/analog/anicateg/can/)

It can be difficult to choose one. After considering several options, we have decided to use Microchip's MPC2510.

Since MPC2510 converts the OBDII CAN bus only into a microprocessor compatible bus (still not compatible with the RS232 found on ordinary PCs), there is one more conversion necessary, which can be solved by using one of the many available microprocessors (one could reached after the PIC microprocessor also manufactured by Microchip).

IN LAYMAN'S TERMS: The CAN interface (not again!) is not directly compatible with standard PCs and requires many months of professional development to create (and is a pain in the wallet). Unless you are crazy, just buy the finished tool for a couple of hundred bucks.

An example of an OBDII message

The diagnostic scanner and the vehicle communicate with each other through exchanging small pieces of information, called diagnostic messages. The message transmitted from the scanner to the PCM is called a command. The message transmitted back from the PCM to the scanner is called a response.

Once the OBDII scanner tool is connected, it will transmit a command message to the vehicle. The PCM microprocessor will receive the message and verify it. If the message is valid, the PCM will respond.

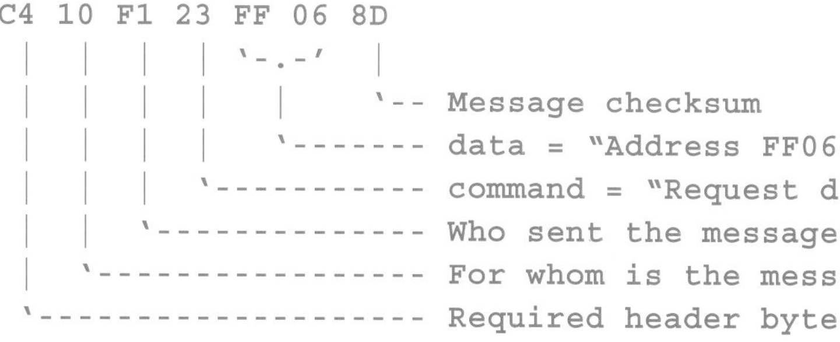

A typical scanner command looks as follows: In the message above, the scanner is requesting data from the PCM memory at address FF06.

The first three bytes are always the same. The fourth byte is the command byte (the PCM response will be based on this command byte). The last byte is a safety measure, so called checksum. The PCM uses the checksum byte to verify that the entire message is not corrupt. All the remaining bytes are the actual data bytes, carrying most of the information (for example the coolant temperature or the address of memory).

The PCM should respond with a similar

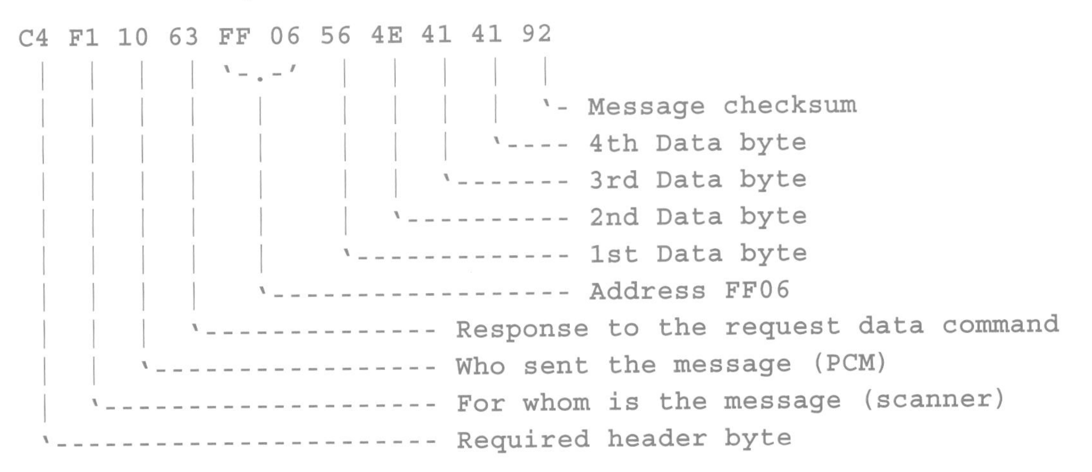

message (data bytes will vary):
The scanner can transmit many kinds of messages, described in the OBDII literature. If an invalid message is transmitted to the PCM, it will simply ignore it.

By now you should have a pretty good understanding of OBDII. Now, what to do with it?

Well, OBDII is meant to help you diagnose a car, but you will need a Scan Tool. In this chapter, you will learn the different types of scan tools which are out there, and what to expect of each.

Chapter 6 > OBDII Scan Tools

Introduction

Today, the technician cannot just listen to an engine and repair it as in the past. If we were to remove a spark plug wire, or coil pack, the engine management will adapt to that missing cylinder. Diagnosis has become much more complex, because manufacturers had to design a vehicle complying with the standards of OBDII.

IN LAYMAN'S TERMS, in order to sell in the United States, manufacturers had to design a new, more complex vehicle around OBDII.

In the past, manufacturers added a few parts to an existing vehicle to comply with emissions. If an emission related part failed, the mechanic would just remove it. I remember seeing mechanics (of course I never did anything like this...) I worked with: gutting CATs, filling EGR valves with lead, grounding O2 sensor control wires to get the vehicle to run richer, putting resistors in the ETC wire to get the vehicle to run richer, and so on. The funny thing is that the vehicle would run fine, because these 'fixes' were just restoring the vehicle to its pre-OBD state.

Today, the vehicles are designed for OBDII. That means they are made to work with this system. If a mechanic were to 'fix' the defective parts, the vehicle would not run properly. The mechanic of yesterday MUST become a technician today. A mechanic will have knowledge of how to repair a vehicle (and usually a great knowledge). The technician of today must posses the knowledge of computer science plus mechanical knowledge. The technician must know how to read waveforms from an oscilloscope; how data buses work; how to read complex wiring diagrams; etc.

IN LAYMEN'S TERMS, OBDII is responsible for the forward movement in automotive technology.

Now let's list the types of scan tools that are on the market.

Tool types

Stand-alone handheld scanners

Stand-alone handheld scanners are the most popular, and the oldest type out there. These are the scanners that you can hold in your hand, and don't need any assistance from a PC, or any other outside source in order to diagnose a car. All they need is your hand, and a car to hook up to. Most use the power source from the vehicle. Very few use an on board battery source. Batteries make it heavy, and drive up the cost of the tool. The cost of most tools are from $350 to $900, depending on the capabilities they have. If a tool is more that $500, be careful about what it does to warrant that cost.

The key features to look for on stand-alone tools are:

1. How easy is it to manipulate the tool? In other words, what feature does it take to move the tool about from screen to screen? Does it have buttons, scroll bar, joystick... and are they easy to use?

2. What kind of display does it have? Don't worry about how big the display is. Worry about if you can read what the tool says. I have seen some tools that have 20 rows of text, and it is so small that it is useless. Make sure you can read the tool.

3. How long is the cable going to the DLC? I have seen some tools where the cable is only three feet long. With the position I was in, while on a road test trying to read the live data, people in the other cars were thinking I was doing unmentionable things to myself. So, make sure the cable is long enough to properly read the tool. By the way, if you plan to go on a road test, and read the live data, please have someone go with you to either drive, or read the tool. Also, make sure the cable is long enough to reach the engine compartment, in order to do tests on the engine.

4. Check out the Ergonomics of the tool. Make sure that the tool is ergonomically correct for you. Ask yourself: Does it fit in your hand? Is it too heavy? Do you need two hands to use the tool?

5. Is the tool upgradeable? Just because the tool is called stand-alone, doesn't mean that it can't be updated. See if the tool can be connected to a PC, to update. Also, see if you can print from it. It is nice to show your customer what is wrong with their vehicle.

These are just some of the questions that you should be asking yourself before buying a tool. Listed below are some scan tools that are on the market.

PDA based

Ok, first what is a PDA? PDA means Personal Digital Assistant. How do you get OBDII to a PDA, and why would you use one? First, the PDA is a handy tool for most shops. The service adviser can: take the information about a customer, do a diagnostic check for faults, and check the history of the customer. Then, he or she can send all this information (via RF) to their PC stations and, by the time both the customer and the adviser get there, all is done. Pretty slick, huh?

Now you may be are asking: I am not a shop, why do I need one? Cost. That is the only reason why a back yard mechanic would need one. If you already have a PDA and it is compatible with the OBDII software out there, go for it. It is one of the cheapest alternatives out on the market. I have seen prices start at $250 and wind up at $500. If any of these are above $400, be careful as to what will warrant that cost. Please remember, these prices are without a PDA.

There are only two operating systems that these will work on, they are: Palm OS, and Windows CE. The Palm is the most reliable that I have seen.

The PDA's that I have seen them work on are:

- Palm m100, m105
- Palm III series
- Palm V series
- Palm VII series
- Handspring Visor, Visor Deluxe, Visor Platinum, Visor Prism, Visor Neo, Visor Pro.
- Handspring Visor Edge.

Most claim to work on the Sony CLIE, but I have yet to see it. Please check all information, since technology may change after this book is published.

You are probably thinking: "that is the way to go", but... Be careful. Nothing is perfect. The problems are:

- Drop one, and it is a goner. They sell covers that make it drop proof, but they look really weird, and are very bulky.
- Most companies that make this software for the PDA's will claim that they work on All PDA's but check the fine print. Most will not work because of the adapter to the car. Adapters and special cables are the problems that most customers are having. Some will sell you

the software, but neglect to tell you, you need a $45 cable or adapter. Others will say: "Oh, you want the to work option... Well that is extra." So, again, be careful.

Laptop based or PC based

Laptop or PC based software is very similar to the PDA. It is a software that will install on the machine, and will have an interface to the vehicle. Prices are around the same for the PDA and laptop, PC. The good thing about laptop or PC based software is that you have a large screen to view all the information you need, without having to scroll through the screens. Also, laptops or PCs are much faster than any machine you can buy.

Once the software is on the machine, most use the serial port to communicate to a vehicle. I have seen some that use the parallel port, but the majority use the serial port. I haven't seen any that use the USB port. Once you are connected to the PC side, then you are supplied with a cable and connector that will communicate to the vehicle. The box that is connected to the DLC usually has some hardware that will decipher the information so the laptop or PC can understand the packets of information being sent.

Now, for the problems:

- The problems with these are the mobility of the tool. A PC is stuck in one place. The laptop is moveable, but it is very bulky.

- Most new laptops don't come with serial ports, so check it out.

- Most of the software sold doesn't work on certain laptops. So, check with the vendor. Make sure before you buy anything, that the software will be compatible with your machine.

- As with the PDA, drop these suckers, and you have a mess, plus you are out a lot of money. Dropping is never covered under warranty.

- So, for practical purposes, this is not the best way to go for OBDII.

OBDII Scan Tool Requirements

With the complexity of the vehicles of today, the use of a scan tool is now mandatory. Not all scan tools are created equal. Before the technician invests in a scan tool, he must look at what the tool will do. Most technicians are tool junkies (myself included), and cost is usually not an consideration. However, with the OBDII standards, all these tools should do the same things, so cost is the number one feature that the consumer will look at.

We are going to look at all the features that an OBDII scan tool is required to have.

Once you have picked the type of tool you want, make sure that it does all that you want it to do. Even though OBDII is a standard that all tools supposed to follow, they don't always do it completely.

According to the SAE standards, the scan tool should provide compatibility with OBDII equipped vehicles. As mentioned in chapter 3, SAE has developed guidelines to facilitate the standardization of OBDII vehicle technology, information and equipment. The standardizations that were discussed were for the vehicle manufacturers to follow. SAE also developed standardizations that the makers of scan tools must follow. These standardizations are as follows:

Connector

The tool must have the standardized 16-pin trapezoidal connector described in SAE J1962. This is so the tool can connect to the DLC of the vehicle.

Basic Functions

The tool must support the four SAE J1978 basic functions:

1. Automatic hands-off determination of the communication protocol (refer to chapter 4 for protocol explanation). Basically this means that the tool must recognize which protocol the vehicle is using, but it has to do it without the user choosing the protocol. Thank goodness for that; I would not have a clue which protocol to choose.

2. Obtaining and displaying the status and results of vehicle on-board diagnostic evaluations (supported and completed readiness tests and malfunction indicator lamp [MIL] status)

3. Obtaining and displaying diagnostic trouble codes (DTCs). Now this is where most tools will differ. The first difference is how the tool displays faults.

Does the tool get the faults with just the fault number, or do you get the number and the fault text? This is an important question to ask. You don't want to pay premium price for a tool if it just gives you a fault number with no text. Also, once the tool gives you the fault number and text, which fault codes do you get the text for? There are two types of fault retrieval:

- **Standard OBDII Fault Retrieval**

Almost all tools that give you a number and text only give you the P0xxx (or Generic) fault text. They won't give you the P1xxx (Manufacturer specific) code text.

- **Extended OBDII Fault Retrieval**

(not all tools comply!)
This type of tool will give you the P0xxx and P1xxx fault and text. This is the tool to get. Make sure that you ask the tool sales person if the tool will do this. Once you get the tool, test it to make sure that it will accomplish this task.

4. Clearing stored emissions related DTCs, freeze frame data and diagnostic test results. Most think that when you clear a fault, the light will go out, and the fault is erased. Well, when you clear a fault, you also clear the freeze frame and all the monitoring statuses.

Freeze Frame Data (not all tools comply!)

The tool should have the capability to check the emissions related freeze frame data.

This is very important to the diagnosis of an engine. What Freeze Frame means is that when a fault is retrieved, helpful data is stored. This data is usually what the engine RPM, temperature, fuel trim, open/closed loop, and others, was at when the fault happened. The engine control unit will take a picture of this data and store it for a tool to review later. No need to explain how important this is!

Real Time Data (not all tools comply!)

The tool should have the capability to check emissions related current data (so called Real Time data, or Live Data).

Real Time Data will tell you what the engine is doing at the time it is doing it. If you suspect the throttle position sensor is failing, go to the Real Time Data, and look at what the TPS is doing. If the engine is at idle, and the tool displays full throttle, then you know you have a problem.

Internal Functions of the Scan Tool (Modes and PIDs)

In order for the tool to perform the functions that were listed above, SAE made up a standard that is called SAE J1979. SAE J1979 describes diagnostic test modes for emission related diagnostic data that is displayed by all scan tools. Ok, that sentence was taken from a textbook. What does this really mean to me, and how does this help me fix a car? Quite simply it won't. SAE J1979 is a standard that is used when you use the tool, and you don't even know it is happening. However, in order for this to be a complete OBDII book, it is necessary to explain what all this is. The SAE J1979 standard is required so that all the manufacturers will follow the same procedure for tool operation. Meaning, when you press the Get Faults button on the tool, it must do the same procedure whether you have a $4000 MasterTech, or a $399 Hellion.

The tool and the car communicate using commands and responses. The tool will send a command, and the car will respond. J1979 sets the standard for the commands and responses. Each command that the tool sends is requesting a test question. Each response tells the tool the answer to the test question.

The test according to J1979 is called a Mode. For example, if the tool sends a request to the engine control module to Get Faults, the command and response will look like this:

Command:

```
68 6A F1 03
         |
         '--- Mode number
              (see later)
```

Response:

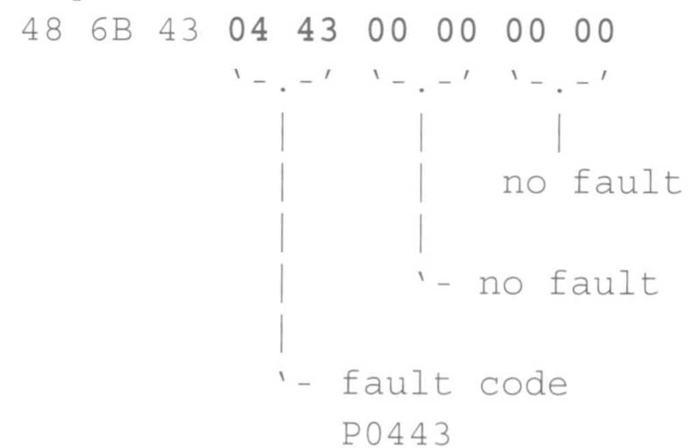

In addition to the Mode number, there can be a Parameter Identification (or a PID). Mode number 01 has many tasks. Therefore, J1979 adds one more digit to the Mode, which is the

PID. If you request the scan tool to display the Engine Coolant Temperature, then the command and response will look like this:

Command:

```
68 6A F1 01 05
         |  |
         |  '- 05 = PID
         '---- 01 = Mode
```

Response:

```
48 6B 06 41 05 50
               |
               '- PID response
```

IN LAYMAN'S TERMS, in order to find out what PID response 50 is, you must look it up in a table in the SAE J1979 standard. That will also explain the conversion of any PID response to real world values.

List of J1979 Modes and PIDs

Mode #01

Request for current powertrain diagnostic data

PID 00 = Lists all the PIDs that are supported by the control module.
PID 01 = Lists the number DTCs, MIL status, all Components being monitored and tells you if they are supported, and the status of each component
PID 03 = Lists the Fuel System status. Meaning, if the fuel system is in Closed Loop, or if it is in Open Loop mode.
PID 04 = Lists the Live Data value for Calculated Load
PID 05 = Lists the Live Data value for Engine Coolant Temperature
PID 06 = Lists the Live Data value for Short term fuel trim Bank 1
PID 07 = Lists the Live Data value for Long term fuel trim Bank 1
PID 08 = Lists the Live Data value for Short term fuel trim Bank 2
PID 09 = Lists the Live Data value for Long term fuel trim Bank 2
PID 0A = Lists the Live Data value for Fuel pressure
PID 0B = Lists the Live Data value for Intake manifold absolute pressure
PID 0C = Lists the Live Data value for

Engine RPM
PID 0D = Lists the Live Data value for Vehicle Speed
PID 0E = Lists the Live Data value for Ignition timing advance for Cyl. 1
PID 0F = Lists the Live Data value for Intake air temperature
PID 10 = Lists the Live Data value forAir flow rate from MAF sensor
PID 11 = Lists the Live Data value for Absolute throttle position sensor
PID 12 = Lists the Live Data value for Secondary air status
PID 13 = Lists the Locations of the Oxygen sensors
PID 14 = Lists the Live Data value for Oxygen sensor 1 Bank 1
PID 15 = Lists the Live Data value for Oxygen sensor 2 Bank 1
PID 16 = Lists the Live Data value for Oxygen sensor 1 Bank 2
PID 17 = Lists the Live Data value for Oxygen sensor 2 Bank 2
PID 18 = Lists the Live Data value for Oxygen sensor 1 Bank 3
PID 19 = Lists the Live Data value for Oxygen sensor 2 Bank 3
PID 1A = Lists the Live Data value for Oxygen sensor 1 Bank 4
PID 1B = Lists the Live Data value for Oxygen sensor 2 Bank 4
PID 1C = Lists the OBDII requirements to which the vehicle is designed for.
PID 1D = Lists the Locations of the Alternate Oxygen sensors

Mode #02
Request for powertrain freeze frame data

PID 00 = Lists all the PIDs that are supported by the control module.
PID 02 = Lists the DTC that caused the Freeze Frame Data
PID 03 = Lists the Fuel System status. Meaning, if the fuel system is in Closed Loop, or if it is in Open Loop mode.
PID 04 = Lists the Freeze Frame Data value for Calculated Load
PID 05 = Lists the Freeze Frame Data value for Engine Coolant Temperature
PID 06 = Lists the Freeze Frame Data value for Short term fuel trim Bank 1
PID 07 = Lists the Freeze Frame Data value for Long term fuel trim Bank 1
PID 08 = Lists the Freeze Frame Data value for Short term fuel trim Bank 2
PID 09 = Lists the Freeze Frame Data value for Long term fuel trim Bank 2
PID 0A = Lists the Freeze Frame Data value for Fuel pressure
PID 0B = Lists the Freeze Frame Data

value for Intake manifold absolute pressure
PID 0C = Lists the Freeze Frame Data value for Engine RPM
PID 0D = Lists the Freeze Frame Data value for Vehicle Speed

Mode #03
Request emission-related powertrain diagnostic trouble codes (DTCs)

This should be a two-step process for the scan tool.

First, the tool sends the command with Mode 01 and PID 01 to request the number of faults that are stored in the control unit. This is done to notify the tool how many faults it should receive.

Second, the tool sends the command with Mode 03. This is to request the fault code numbers.

Mode #04
Clear/Reset emission-related diagnostic information

- Clearing the number of DTCs
- Clearing the DTCs
- Clearing the trouble code for Freeze Frame Data
- Clearing the Freeze Frame Data
- Clearing the Oxygen sensor test data
- Testing the status of the system monitoring tests
- Clearing all on-board monitoring test results

Mode #05
Request Oxygen sensor monitor test results

The purpose of this Mode is to allow access to the on-board Oxygen sensor monitoring test results. This Mode is optional because most vehicle manufacturers don't support this requirement. Therefore, if you are using the tool and the tool does not respond or acts strangely, note that the Mode might not be supported.

Mode #06
Request latest on-board monitoring test results for Non-Continuous Monitor systems

- Catalyst
- Exhaust gas re-circulation (EGR)
- Evaporative system

Mode #07
Request latest on-board monitoring test results for Continuous Monitor systems

- Fuel trim
- Misfire
- Comprehensive components

Mode #08
Request Control of On-Board System, Test, or Component

The purpose of this Mode is to enable the off-board test device to control the operation of an on-board system, test, or component.

Mode # 09
Request Vehicle Information

This Mode is to retrieve the Vehicles VIN, and calibration IDs. This Mode is not always supported by the manufacturer.

(I'll bet you never knew all this was happening behind the scenes of your scan tool!)

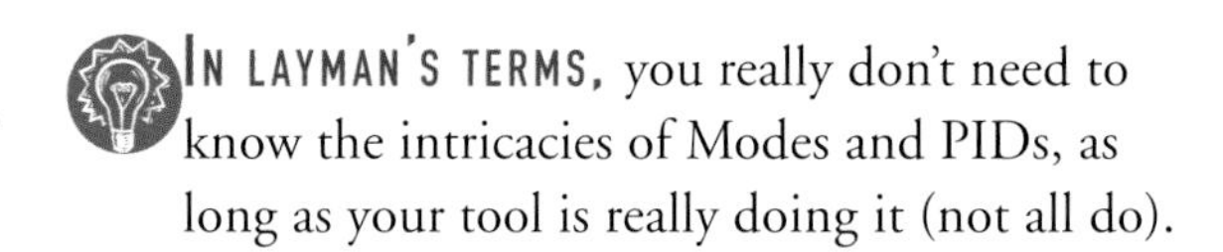

IN LAYMAN'S TERMS, you really don't need to know the intricacies of Modes and PIDs, as long as your tool is really doing it (not all do).

True OBDII compatibility Checklist

At this point, you have selected the type of tool, and you know what it should do. If you are looking for a tool with true OBDII compatibility, before you put your money on the table, use our checklist:

1. Is the tool able to check or monitor the status of the following components?

- Catalyst
- Heated Catalyst
- Misfire and Misfire for Diesels
- Evaporative System
- Secondary Air System
- Air Conditioning System Refrigerant
- Fuel System
- Oxygen System
- Exhaust Gas Recirculation (EGR) System
- Positive Crankcase Ventilation (PCV) System

- Thermostat
- Comprehensive

In addition to checking status, is the tool able to check if these systems are supported or not? Some vehicles will not support all these systems.

2. Does the tool support Extended OBDII Fault Retrieval?
(For details, see the section Basic Functions above).

3. Does the tool reset OBDII Faults?

4. Does the tool display Freeze Frame Data?

5. Does the tool display Oxygen Sensor Inputs?
This is helpful for checking if the Oxygen Sensors are supported by OBDII. If the tool is fully compatible, it will display the voltage, switching time, and max/min values. Most vehicles before 2001 will not completely support this function. If the vehicle supports this function, then this is very useful to diagnosis an engine. You can learn a lot from an Oxygen Sensor. You can see how the engine is running by how it is switching.

6. Does the tool supply Live Data?
Most technicians will use the Live Data more than any other function of the scan tool.

7. Does the tool display VIN?
Vehicle Identification Number input is not important, but helpful. This will display the VIN of the vehicle. Most vehicles produced before 2001 will not support this function.

8. Does the tool allow input of Manual Commands?
Manual Commands is another function that is helpful. What Manual Commands does is allow the scan tool owner to enter and send the Mode number or PID manually. The Mode number, PID and the response are displayed in hexadecimal.

As new vehicles are developed with new Modes and PIDs (not yet supported by your current tool), this feature helps your tool not to become obsolete.

Closing thoughts for purchasing a Scan Tool

Well, there you have it. These are the basic types and capabilities of the OBDII scan tools. We listed some of the pros and cons for each.

In our opinion:

> *For professional use,* the handheld tester is the best way to go. Take all that we said about these types, and ask yourself this question: "Will this tool make money for me in the long run?" Obviously, money is a consideration, but the ease of use and the amount of information the tool will give you can be the more important.
>
> *For the do-it-yourselfer,* the tool is not meant to make you money. It might be used only three times a year, so price can be is the main consideration. If you are a weekend warrior only, and own a PDA, you might consider the PDA software.

I can't say it enough, DO YOUR HOMEWORK, check out every aspect. There is a lot of competition out there, so be careful and wise in your selection.

Happy hunting!

Ok, now that you are armed and ready with your scan tool, what are you going to do with it?

This chapter discusses the basics on how to infiltrate engines with your scan tool.
You will see how to interrogate the engine for faults and live data, and how freeze frame data will help you with the whole picture and lots more.

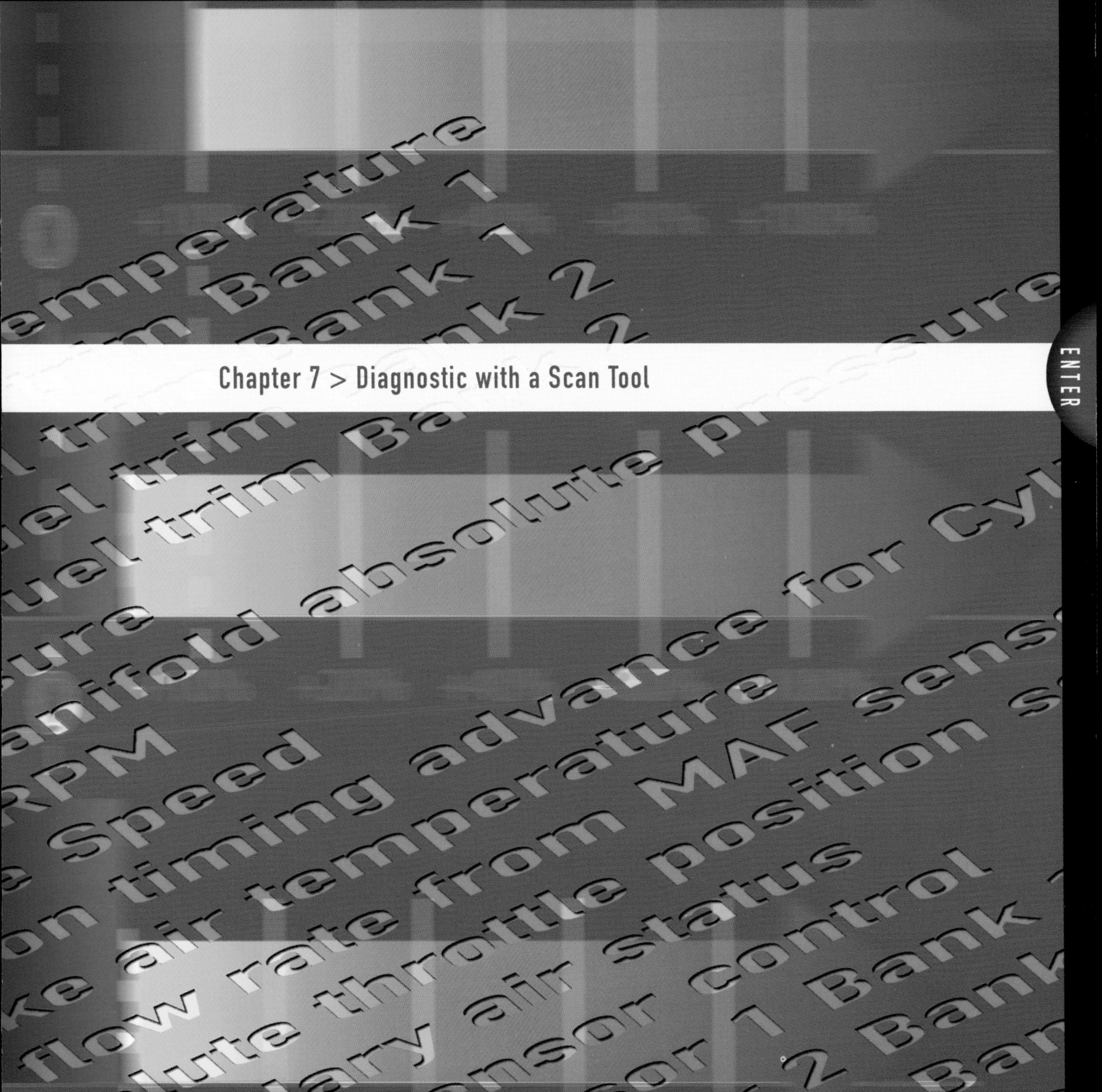
Chapter 7 > Diagnostic with a Scan Tool
ENTER

Introduction

In the previous chapter, we looked at the kinds of scan tools that are on the market. You have an idea of the tool you want, but, what will you do with it? We will look at all the functions that the scan tool must do, and put them to practical use. We will be using the Hellion handheld tool, manufactured by DiabloSport (you can check it out at www.eobd.com).

Now, we will break down all the functions which an OBDII scan tool performs.

General Data

First, we will look at how to use Component monitoring. In the chapter The Anatomy of OBDII, we briefly explained what Component monitors do. Here is what to do with them

Component Monitoring

This section lists all Component Monitors and their classifications. Before you check the status of a monitor, check if the monitor is supported. If not, you can look at the status all day, and not see a response (that wouldn't be too smart, huh?). Don't laugh - we have seen this done before.

Continuous Monitors

The first type of a Monitor that we will discuss is the Continuous Monitor. Continuous Monitors are, well, do I need to say, they are Monitors that are being done continuously. Once the engine reaches the proper conditions, the Monitors start their tests and continue until the engine is turned off.

Misfire and Misfire for Diesels

This Monitor is probably the most important and most complicated of all the Monitors. The Misfire Monitor checks if there is a misfire. Unlike the old days, when a misfire was blamed only on the ignition system, this monitor checks any condition that can cause a misfire. Misfires can come from (of course) the ignition system, vacuum leaks, sticking EGR valve, an air/fuel ratio imbalance, PCV valve sticking, or faulty, incorrect timing (due to a worn timing belt), fuel injectors sticking, or leaking, and so on. To put it simply, anything that has to do with a problem with the combustion efficiency, can create a misfire.

The Misfire is detected by the crankshaft sensor. This sensor is used to sense the speed the engine is traveling. In addition, it is used to determine if there has been a misfire in the engine. This is done by measuring the speed of each combustion stroke. The PCM knows how much thrust each stroke produces, and if one of these strokes uses less thrust, then it is determined that there has been a misfire.

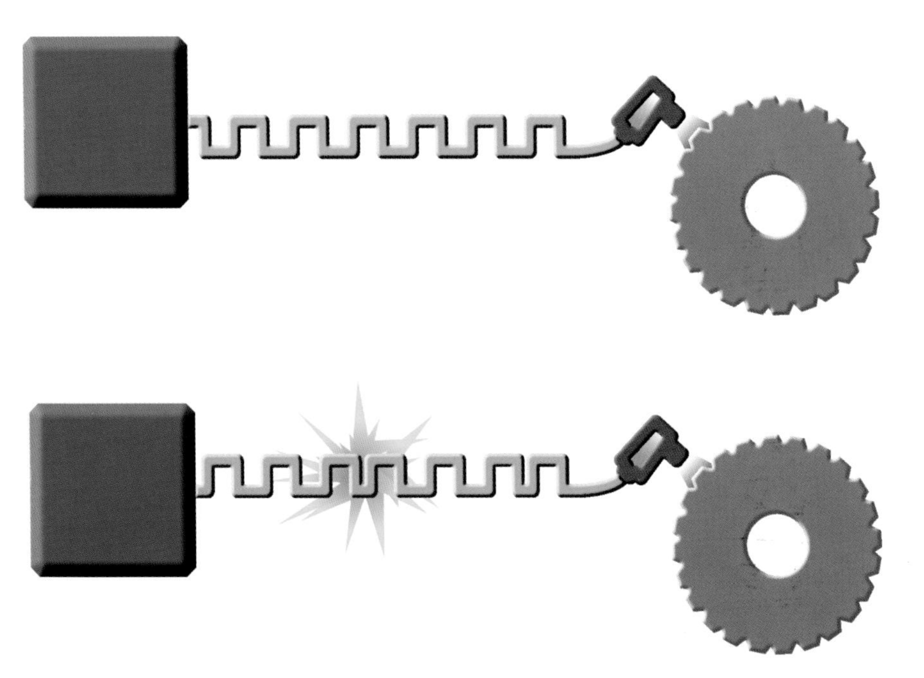

If a misfire is detected, then you will see the famous P0300 fault code. If a misfire can be pinpointed to a specific cylinder, then you will get a P03XX. The XX identifies the cylinder that has caused the misfire. For an example, if you get a P0302, then cylinder number 2 is the problem.

Fuel System Monitor

The Fuel Systems Monitor is an engine control unit diagnostic that monitors the adaptive fuel control system. To understand what all this means, we have to look at Long Term Fuel Trim (or LTFT) and Short Term Fuel Trim (or STFT). STFT is what keeps the engine running at the best overall air/fuel ratio of 14.7:1. This air/fuel ratio is the "Perfect" running engine, and is called Stoichiometry.

STFT is a control unit parameter that is used to indicate short-term fuel adjustments. This parameter is expressed by the scan tool as a percentage and its range is from -10% to +10%. LTFT is determined by the STFT. LTFT takes samples of the STFT to see if it needs to shift, so STFT can stay at its optimum range. Below is an example of Fuel Trim.

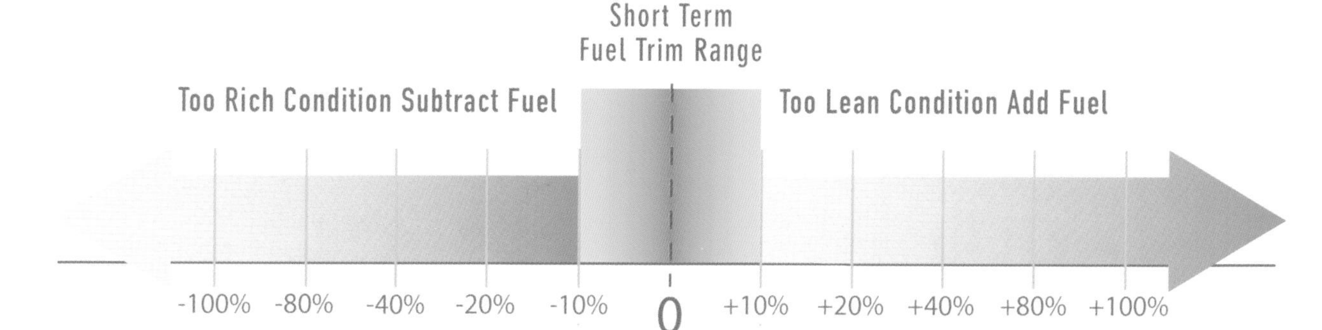

To gain a better understanding, lets take an example. Take an engine that has reached the closed loop, and the control unit has received a signal from the Oxygen Sensor that indicates the air/fuel mixture is richer than desired. The control unit will move the STFT to a more negative range to compensate for the rich condition. If the control unit sees that the STFT has been adjusting for a rich condition for a long period of time, the control unit will Learn (or Adapt) to this condition. This will move the LTFT. The LTFT will shift the whole scale into a negative range to compensate so the STFT will stay at a value close to 0%.

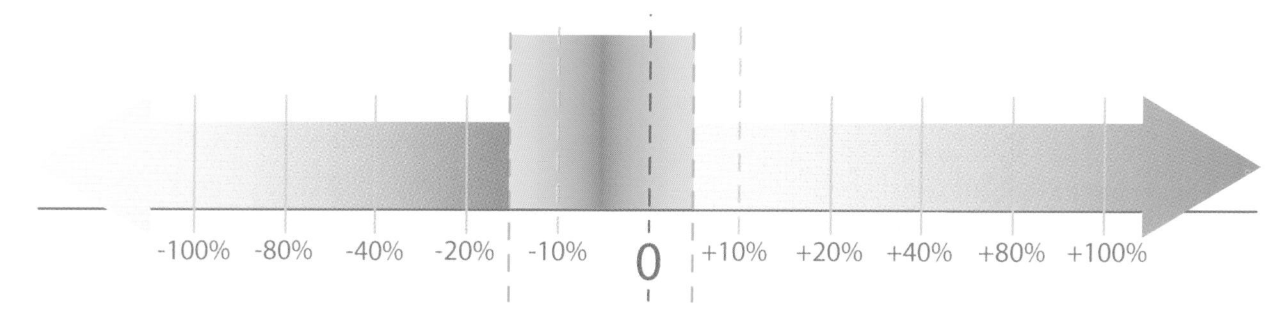

Once any change happens, whether with STFT or LTFT, the control unit adds a correction factor to the injector pulse width calculation to adjust for variations. If the change is too large, the control unit will throw a DTC.

STFT values are not permanent; they are erased as soon as you turn off the ignition switch. Each time you start the engine, the STFT starts over. LTFT is stored in the Keep Alive Memory (KAM). You will need a scan tool to reset this value. Once you reset faults, this resets the KAM.

Comprehensive Component Monitoring

This is a diagnostic system test of engine components that are not continuously monitored though any other OBDII test. These components must have an input to the control unit, and if they fail they must have an affect on the running of the engine. The components are as follows:

Input Components:

- Vehicle Speed Sensor
- Crank angle sensor
- Throttle position sensor
- Coolant temperature sensor
- Cam position sensor
- Fuel composition sensor
- Intake air temperature sensor
- MAF sensor
- MAP sensor
- Knock sensor
- Brake switch
- Gear selector switch
- Cruise control switch

Output Components:

- Automatic idle speed motor
- Emission related electronic only transmission controls
- Heated fuel preparation systems
- Glow lamp (on Diesels)
- Warm up catalyst bypass valve

Non-Continuous Monitors

Next, we have the Non-Continuous monitors. These monitors will run until their tests have been satisfied. Then they will stop. So, check with the scan tool to see if the monitors are done, and, if not, continue to check the tool to see when they complete. Here is a list of all the Non-Continuous Monitors:

Catalyst Monitor

This monitor checks the performance of the Catalytic converter. This is accomplished by comparing the Oxygen sensor before the CAT, and after the CAT. The sensor before the CAT usually is the sensor that is the most active. If the sensor after the CAT is very active also, then the monitor will fail this test.

Heated Catalyst Monitor

This is a monitor that checks the efficiency of the way the CAT heats up. If this monitor is not complete, you need to check the Oxygen sensor heaters, air injection system and the CATs themselves.

Evaporative System Monitor

This monitor checks the EVAP system. The EVAP system is the system that controls fuel vapor, in the fuel tank, and the fuel return. All EVAP systems are different, so each will be covered in their perspective chapter for diagnostics. Loose, or missing gas caps are the number one reason why the MIL lamp is on. Therefore, if you have an EVAP fault, and the system has failed, then your first check should be the gas cap.

Secondary Air System Monitor

This monitor checks the air pump, and all its components. The importance of this monitor is to check how much air is being pumped to the CAT during warm up. This system will pump air to the CATs to have them fire up faster in order to burn off the excess fuel in the exhaust.

Oxygen Sensor System Monitor

This monitor is one of the most important of all the monitors. The Oxygen Sensor is the heart and soul of the fuel system. It tells the control unit how much fuel is being burned or not burned. It also is the main feed back for both Short Term, and Long Term Fuel Trim.

Exhaust Gas Recirculation (EGR) System

This monitor checks the EGR valve, and the EGR flow. Most of the common problems with the EGR is that the flow pipes clog up with soot. For a more complete description of the ERG system, please refer to each manufacturer section for diagnosis.

Positive Crankcase Ventilation (PCV) System Monitor

Not much to explain about this one - if this monitor fails, check the PCV valve, or the pipes that are used.

Air Conditioning System Refrigerant Monitor

This monitor is the only one that I have not seen a manufacturer support. It is supposed to check the idle increase with compressor engagement. In addition, it checks engine torque when the compressor kicks in.

Thermostat Monitor

The only vehicle that I have seen use this monitor is BMW. BMW monitors the flow of the coolant system, since if the thermostat breaks, or is stuck (which happens often) the ECT sensor will not work efficiently.

All the monitors are important to guide the technician to the area of the fault. Yes, getting the fault is an important task, but the Monitoring tests tell you what to look at.

For instance, if you have a miss in the engine, the DTC P0304 is set. You might look at cylinder number 4 as being the culprit. However, instead of diving into an injector test or ignition test, check the systems monitor when you see the Exhaust Gas Recirculation system is not complete. You might look at the EGR valve being stuck as a cause for a vacuum leak at idle.

Therefore, faults are not the only information that will help you with the diagnosis of the vehicle. Check out all the aspects of the engine with your scan tool. You will see that systems monitoring is a crucial part of the diagnosis.

Fuel System Status

Use this selection to check the condition of the Fuel System. You are checking to see if the system is in Closed Loop Mode, or in Open Loop Mode.

Secondary Air System

A lot of vehicles don't have this system anymore. Use this to check if the vehicle has a Secondary Air System

Oxygen Sensor Locations

I like to use this to check which Oxygen Sensors are located on a vehicle. On some vehicles, the Oxygen Sensors are not always easy to see. The possible locations are as follows:

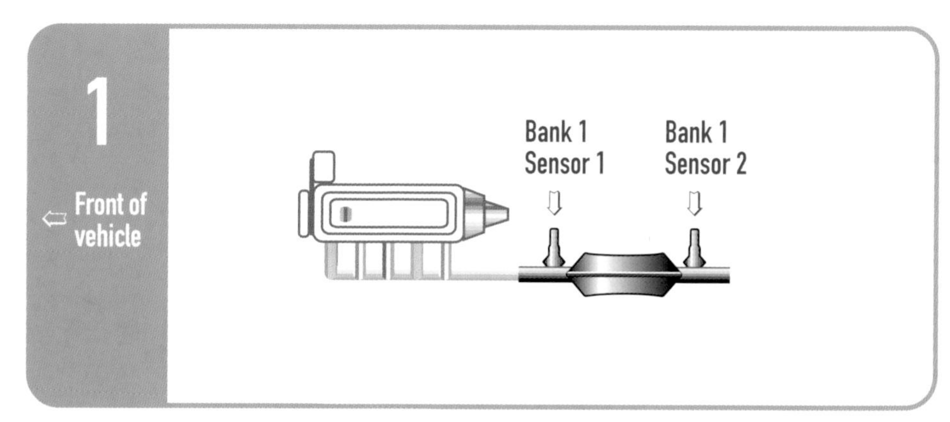

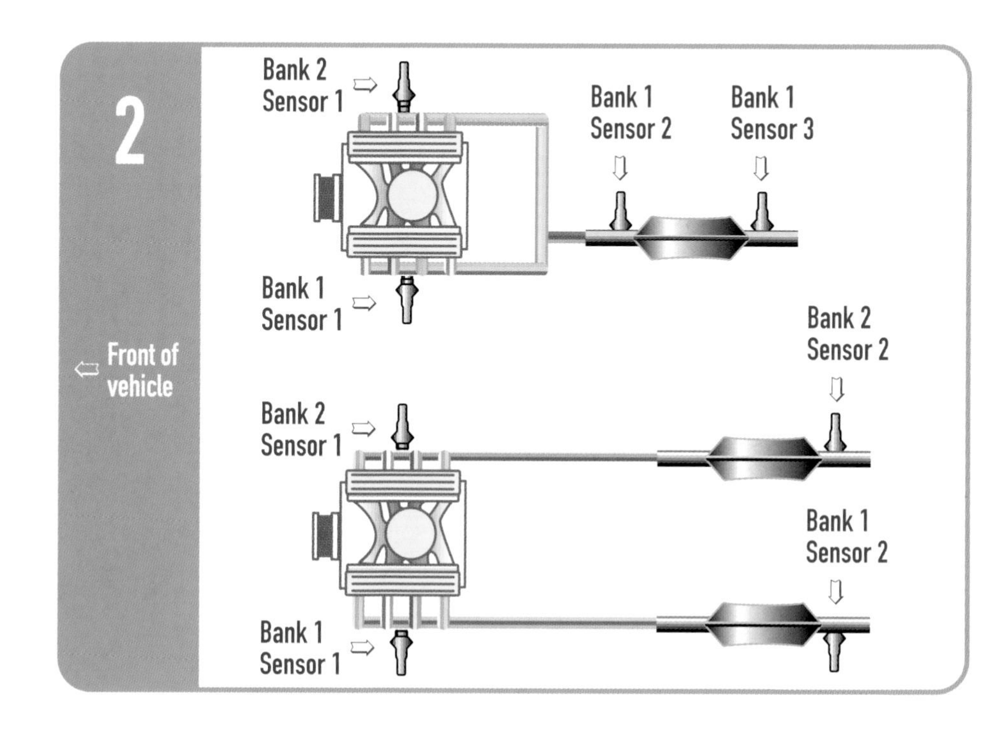

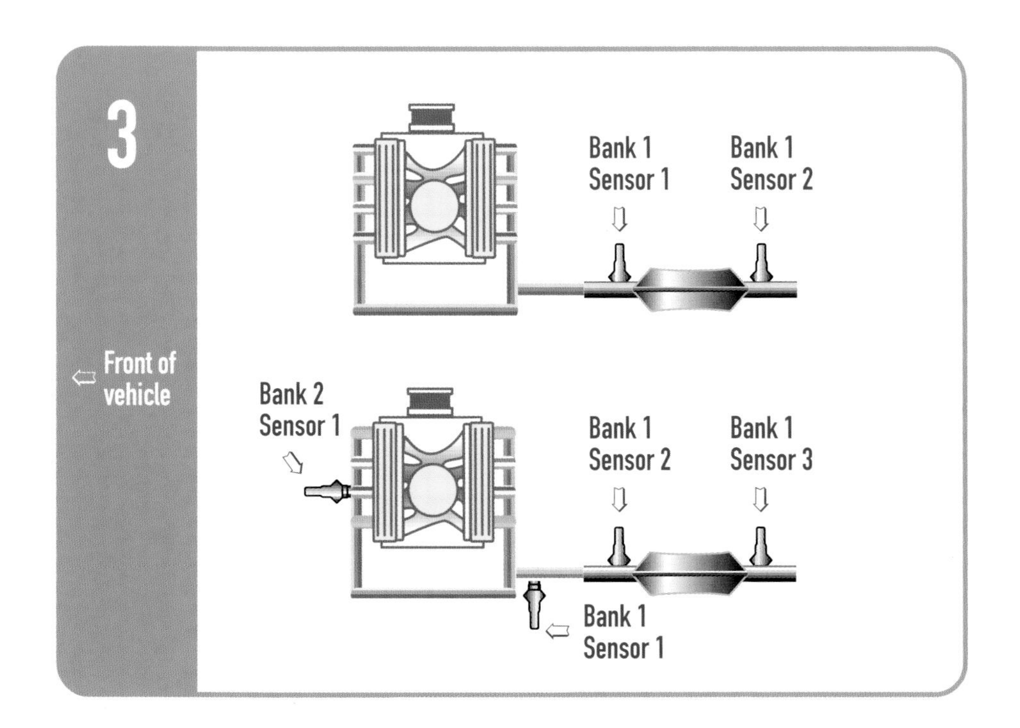

Inputs Supported

This will list all the Real Time Data values that the vehicle will support.

MIL Status

Use this selection to check it see if the MIL is listed as ON or OFF. It is a good check to see if someone has removed the MIL bulb, or that the circuit is broken.

Trouble Codes

Getting faults is most likely the first part of using the scan tool. However, it can be the most deceiving part of repairing the vehicle. Just because a fault is listed, doesn't mean that the component listed in the fault is defective. If you have the fault of P0130 (O2 Sensor Circuit Malfunction, Bank 1 Sensor 1). That doesn't mean that the Oxygen sensor is bad.

You have to check out the whole system before condemning the part. I have seen many mechanics make that mistake; condemn a part just because the tool tells them to.

Lets get something straight right now, the scan tool is not a mechanic in a plastic case, it is not the end-all, fix-all. The tool is only a guide to show you where to look for the problem.

Ok, now I've gotten that off my chest, back to the Oxygen Sensor. Now that you have the fault code, check the freeze frame data, then check the Real Time Data for the STFT and LTFT, also look at the voltage control of the Oxygen Sensor. See if it is swinging properly. Then look at the systems monitoring, to see if the monitor is complete or not.

All these checks will help you with the diagnosis. Now, if the STFT, and the LTFT are in the positive range, and the Oxygen Sensor is swinging half way decently, then look for a vacuum leak or a fuel delivery problem. Something has to make the engine have lack of fuel.

Get faults

There are two possible types of faults:

Active faults

Active is when the fault is present, and the MIL is on.

Stored faults

Stored is when the fault has been detected, but the MIL is not on, or was on and went out later.

Remember to write down all the faults, this will help you visualize what is happening. That way you will know what your next step should be.

Reset faults and Freeze Frame Data

Resetting of the faults can be done manually though the scan tool. In addition, it can be done automatically through the control unit. If a fault has caused the MIL to light, and has passed three consecutive monitoring test, on three consecutive road trips, the MIL lamp will go out. Take it one step further. If a fault is not seen in 40 warm ups, and has passed 40 consecutive monitoring tests, the fault will then be erased. Unfortunately, the chances of that happening are slim to none. However, if it ever did, the fault would be erased.

As explained before, resetting of the faults does not only reset the fault, but all other systems too.

Freeze frame data

Freeze Frame data is also important. You can get an idea of what the fault was doing at the time it was logged. The freeze frame data is as follows:

- Fault code number
- If the fuel system is in Closed Loop, or if it is in Open Loop mode.
- Engine RPM
- Vehicle Speed
- Engine Coolant Temperature
- Calculated Load
- Short term fuel trim Bank 1
- Long term fuel trim Bank 1
- Short term fuel trim Bank 2
- Long term fuel trim Bank 2
- Fuel pressure
- Intake manifold absolute pressure

Oxygen Sensor

Use this to check the thresholds of the Oxygen Sensor. You can check the Rich to Lean and the Lean to Rich thresholds. Within each, you will find a Maximum or Minimum values. This is nice to see how well the Oxygen Sensors are responding

Real Time Data

Real Time Data is the second thing that most technicians will check to diagnose an engine. Here is a complete list of all the possible OBDII Real Time Data values (please remember, the PCM of the vehicle will only show supported Real Time Data values):

- Calculated Load
- Engine Coolant Temperature
- Short term fuel trim Bank 1
- Long term fuel trim Bank 1
- Short term fuel trim Bank 2

- Long term fuel trim Bank 2
- Fuel pressure
- Intake manifold absolute pressure
- Engine RPM
- Vehicle Speed
- Ignition timing advance for Cylinder 1
- Intake air temperature
- Air flow rate from MAF sensor
- Absolute throttle position sensor
- Secondary air status
- Oxygen sensor control
- Oxygen sensor 1 Bank 1
- Oxygen sensor 2 Bank 1
- Oxygen sensor 1 Bank 2
- Oxygen sensor 2 Bank 2
- Oxygen sensor 1 Bank 3
- Oxygen sensor 2 Bank 3
- Oxygen sensor 1 Bank 4
- Oxygen sensor 2 Bank 4

Vehicle Information

In this selection, you will find out information about the: VIN, Calibration ID, and the CVN.

Manual Commands

This selection is the least known part of the tool. Most people think that this will do functions that the tool would otherwise not do.

To the contrary, this will do manually all that the tool can already do automatically. If you would like to see what the tool is doing, use this selection as a research project for yourself.

Once you selected the Manual Command on your tool, enter the Mode number, then PID number:

```
Mode = 01
PID  = 01
```

Now Press Enter. The tool executes the command for you (which is why you paid the big bucks for it).

Just look at the response. Your tool is displaying several hexadecimal numbers. What to do with these numbers? Confusing, isn't it?

Getting all those numbers and converting them to meaningful diagnostic information is not an easy task. It is a result of the many hours engineers spent to get the tool to show you the values to diagnose a vehicle successfully (so, when there is an update, don't complain about the price too much...).

With this chapter,
you will take your scan tool
to the next level
for General Motors vehicles.
Even when you know
how to use the tool
to interrogate faults,
each manufacturer
has their own ins and outs.

Chapter 8 > OBDII Diagnostic Tips for GM

General Information about General Motors

GM was the first to initiate the OBDII standards. During the time of OBDI systems, GM used many of the standards that are mandatory for OBDII. The MIL circuit was the same principal as in the OBDII system. Also, they used the Short Term, and Long Term Fuel trim for the engine to adapt to improve driving conditions. The GM OBDI system did not use the system monitoring to check the condition of the emissions.

Understanding General Motors Diagnostics

General Motors uses software called the Diagnostic Executive. This control is how the OBDII systems will be handled. The Diagnostic Executive software is found in the PCM. This is a list of all the tasks of the Diagnostic Executive:

- Controls the Continuous and Non-Continuous Monitors
- Logs the Monitors and displays their status
- Controls the MIL to turn ON or OFF
- Records and clears the DTCs
- Records the Freeze Frame Data
- Controls non-OBDII related MILs

The main function of the Diagnostic Executive is to control the fault code categories, and handle the MIL when a fault has happened. The Diagnostic Executive also controls turning off the MIL. The Diagnostic Executive is the heart and soul of the OBDII system.

General Motors Driving Cycle

The GM drive cycle is a key factor in diagnosis of the OBDII system. When the drive cycle is done, all the tests should have been completed. Have an assistant watch the Readiness status of the Non-Continuous and Continuous Monitors while performing the drive cycle (don't join TAD, the test-and-drive organization).The GM drive cycle has several stages:

Stage 1 Cold Start. In order to be classified as a cold start the engine coolant temperature must be below 122°F and within 11°F of the ambient air temperature at startup. Do not leave the key on prior to the cold start or the heated oxygen sensor diagnostic may not run.

Stage 2 Idle. The engine must be run for two and a half minutes with the air conditioner and rear defroster on. The more electrical load you can apply, the better. This will test the O2 heater, Passive Air, Purge "No Flow", Misfire and, if a closed loop is achieved, Fuel Trim.

Stage 3 Accelerate. Turn off the air conditioner and all the other loads and apply half throttle until 55mph is reached. During this time the Misfire, Fuel Trim, and Purge Flow diagnostics will be performed.

Stage 4 Hold Steady Speed. Hold a steady speed of 55mph for 3 minutes. During this time the O2 response, Air Intrusive, EGR, Purge, Misfire, and Fuel Trim diagnostics will be performed.

Stage 5 Decelerate. Let off the accelerator pedal. Do not shift touch the brake or clutch. It is important to let the vehicle coast along gradually slowing down to 20 mph. During this time the EGR, Purge and Fuel Trim diagnostics will be performed.

Stage 6 Accelerate. Accelerate at 3/4 throttle until 55-60mph. This will perform the same diagnostics as in stage 3.

Stage 7 Hold Steady Speed. Hold a steady speed of 55mph for five minutes. During this time, in addition to the diagnostics performed in stage 4, the catalyst monitor diagnostics will be performed. If the catalyst is marginal or the battery has been disconnected, it may take 5 complete driving cycles to determine the state of the catalyst.

Stage 8 Decelerate. This will perform the same diagnostics as in stage 5. Again, don't press the clutch or brakes or shift gears.

GM specific Fault codes (P1)

Chevy Metro
P1106 - Manifold Absolute Pressure [MAP] Sensor Circuit Intermittent High Voltage
P1107 - Manifold Absolute Pressure [MAP] Sensor Circuit Intermittent Low Voltage
P1111 - Intake Air Temperature [IAT] Sensor Circuit Intermittent High Voltage
P1112 - Intake Air Temperature [IAT] Sensor Circuit Intermittent Low Voltage
P1114 - Engine Coolant Temperature [ECT] Sensor Circuit Intermittent Low Voltage
P1115 - Engine Coolant Temperature [ECT] Sensor Circuit Intermittent High Voltage
P1120 - Throttle Position [TP] Sensor 1 Circuit
P1125 - Accelerator Pedal Position [APP] System
P1133 - HO2S Insufficient Switching Bank 1 Sensor 1
P1134 - HO2S Transition Time Ratio Bank1 Sensor 1
P1153 - HO2S Insufficient Switching Bank 2 Sensor 1
P1154 - HO2S Transition Time Ratio Bank 2 Sensor 1
P1171 - Fuel System Lean during Acceleration
P1187 - Engine Oil Pressure Sensor Circuit Low Voltage
P1188 - Engine Oil Pressure Sensor Circuit High Voltage
P1191 - Intake Air Duct Leak
P1214 - Injection Pump Timing Offset
P1216 - Fuel Solenoid Response Time Too Short
P1217 - Fuel Solenoid Response Time Too Long
P1218 - Injection Pump Calibration Circuit
P1220 - Throttle Position [TP] Sensor 2 Circuit
P1221 - Throttle Position [TP] Sensor 1-2 Correlation
P1258 - Engine Coolant Over-temperature - Protection Mode Active

P1271 - Accelerator Pedal Position [APP] Sensor 1-2 Correlation

P1272 - Accelerator Pedal Position [APP] Sensor 2-3 Correlation

P1273 - Accelerator Pedal Position [APP] Sensor 1-3 Correlation

P1275 - Accelerator Pedal Position [APP] Sensor 1 Circuit

P1276 - Accelerator Pedal Position [APP] Sensor 1 Performance

P1277 - Accelerator Pedal Position [APP] Sensor 1 Circuit Low Voltage

P1278 - Accelerator Pedal Position [APP] Sensor 1 Circuit High Voltage

P1281 - Accelerator Pedal Position [APP] Sensor 2 Performance

P1282 - Accelerator Pedal Position [APP] Sensor 2 Circuit Low Voltage

P1283 - Accelerator Pedal Position [APP] Sensor 2 Circuit High Voltage

P1285 - Accelerator Pedal Position [APP] Sensor 3 Circuit

P1286 - Accelerator Pedal Position [APP] Sensor 3 Performance

P1287 - Accelerator Pedal Position [APP] Sensor 3 Circuit Low Voltage

P1288 - Accelerator Pedal Position [APP] Sensor 3 Circuit High Voltage

P1345 - Crankshaft Position [CKP] Camshaft Position [CMP] Correlation

P1380 - Misfire Detected - Rough Road Data Not Available

P1381 - Misfire Detected - No Communication with Brake Control Module

P1404 - Exhaust Gas Recirculation [EGR] Closed Position Performance

P1406 - Exhaust Gas Recirculation [EGR] Position Sensor Performance

P1408 - EGR Vacuum System Malfunction

P1409 - EGR Vacuum System Leak

P1410 - EGR Vacuum System Restriction

P1415 - Secondary Air Injection [AIR] System Bank 1

P1416 - Secondary Air Injection [AIR] System Bank 2

P1431 - Fuel Level Sensor 2 Performance

P1432 - Fuel Level Sensor 2 Circuit Low Voltage

P1433 - Fuel Level Sensor 2 Circuit High Voltage
P1441 - Evaporative Emission [EVAP] System Flow during Non-Purge
P1508 - IAC System Low RPM
P1509 - IAC System High RPM
P1514 - Throttle Body Performance
P1515 - Commanded vs. Actual TP Performance (PCM)
P1516 - Commanded vs. Actual TP Performance (TAC Module)
P1517 - Throttle Actuator Control [TAC] Module Performance
P1518 - PCM to TAC Module Serial Data Circuit
P1523 - Throttle Closed Position Performance
P1539 - Air Conditioning [A/C] Clutch Feedback Circuit High Voltage
P1545 - Air Conditioning [A/C] Clutch Relay Control Circuit
P1546 - Air Conditioning [A/C] Clutch Feedback Circuit Low Voltage
P1546 - Air Conditioning [A/C] Clutch Relay Control Circuit
P1548 - HVAC System Recirculation Door Motor Control Circuit
P1571 - Traction Control Torque Request Circuit
P1574 - Stop lamp Switch Circuit
P1575 - Extended Travel Brake Switch Circuit High Voltage
P1621 - EEPROM Write Error
P1624 - PCM RAM Performance
P1626 - Theft Deterrent Fuel Enable Signal Not Received
P1627 - Control Module Analog to Digital Performance
P1630 - Theft Deterrent Learn Mode Active
P1631 - Theft Deterrent System - Password Incorrect
P1635 - 5 Volt Reference 1 Circuit
P1639 - 5 Volt Reference 2 Circuit

P1643 - Wait to Start Lamp Control Circuit
P1644 - Traction Control Delivered Torque Output Circuit
P1646 - 5 Volt Reference 3 Circuit
P1652 - Powertrain Induced Chassis Pitch Output Circuit
P1653 - Exhaust Pressure Regulator [EPR] Solenoid Control Circuit
P1654 - Service Throttle Soon Lamp Control Circuit
P1655 - EGR Solenoid Control Circuit
P1656 - Turbocharger Wastegate Solenoid Control Circuit
P1810 - TFP Valve Position Switch Circuit - Invalid Range
P1811 - Maximum Adapt and Long Shift
P1814 - Torque Converter Overstress
P1815 - Transmission Range Switch - Start In Wrong Range
P1816 - TFP Valve Position Switch - Park/Neu. With Drive Ratio
P1817 - TFP Valve Position Switch - Reverse With Drive Ratio
P1818 - TFP Valve Position Switch - Drive Without Drive Ratio
P1842 - 1-2 Shift Solenoid Circuit - Low Voltage
P1843 - 1-2 Shift Solenoid Circuit - High Voltage
P1845 - 2-3 Shift Solenoid Circuit - Low Voltage
P1847 - 2-3 Shift Solenoid Circuit - High Voltage
P1853 - Brake Band Apply Solenoid High Voltage
P1860 - Torque Converter Clutch PWM Solenoid Circuit
P1868 - Transmission Fluid Life
P1870 - Transmission Component Slipping
P1875 - 4WD Low Switch Circuit - Electrical
P1887 - TCC Release Switch Circuit
P1892 - Throttle Position Sensor PWM Signal High

U1000 - Loss of Serial Communications for Class 2
U1026 - Loss of Class 2 Communication with ATC
U1041 - Lost Communications with Brake/Traction Control System
U1042 - Loss of Class 2 Communication with ABS
U1064 - Lost communications with Body Multiple Control System
U1192 - Loss of VTD Communications
U1193 - Loss of VTD Communications
U1300 - Class 2 Data Link Low
U1301 - Class 2 Data Link High
Chevy Prizm
P1107 - Manifold Absolute Pressure [MAP] Sensor Circuit Intermittent Low Voltage
P1111 - Intake Air Temperature [IAT] Sensor Circuit Intermittent High Voltage
P1112 - Intake Air Temperature [IAT] Sensor Circuit Intermittent Low Voltage
P1114 - Engine Coolant Temperature [ECT] Sensor Circuit Intermittent Low Voltage
P1115 - Engine Coolant Temperature [ECT] Sensor Circuit Intermittent High Voltage
P1120 - Throttle Position [TP] Sensor 1 Circuit
P1125 - Accelerator Pedal Position [APP] System
P1133 - HO2S Insufficient Switching Bank 1 Sensor 1
P1134 - HO2S Transition Time Ratio Bank1 Sensor 1
P1153 - HO2S Insufficient Switching Bank 2 Sensor 1
P1154 - HO2S Transition Time Ratio Bank 2 Sensor 1
P1220 - Throttle Position [TP] Sensor 2 Circuit
P1221 - Throttle Position [TP] Sensor 1-2 Correlation
P1258 - Engine Coolant Over-temperature - Protection Mode Active
P1275 - Accelerator Pedal Position [APP] Sensor 1 Circuit
P1276 - Accelerator Pedal Position [APP] Sensor 1 Performance

P1280 - Accelerator Pedal Position [APP] Sensor 2 Circuit
P1281 - Accelerator Pedal Position [APP] Sensor 2 Performance
P1285 - Accelerator Pedal Position [APP] Sensor 3 Circuit
P1286 - Accelerator Pedal Position [APP] Sensor 3 Performance
P1380 - Misfire Detected - Rough Road Data Not Available
P1381 - Misfire Detected - No Communication with Brake Control Module
P1415 - Secondary Air Injection [AIR] System Bank 1
P1416 - Secondary Air Injection [AIR] System Bank 2
P1431 - Fuel Level Sensor 2 Performance
P1432 - Fuel Level Sensor 2 Circuit Low Voltage
P1433 - Fuel Level Sensor 2 Circuit High Voltage
P1441 - Evaporative Emission [EVAP] System Flow during Non-Purge
P1514 - Throttle Body Performance
P1515 - Commanded vs. Actual TP Performance (PCM)
P1516 - Commanded vs. Actual TP Performance (TAC Module)
P1517 - Throttle Actuator Control [TAC] Module Performance
P1518 - PCM To TAC Module Serial Data Circuit
P1539 - Air Conditioning [A/C] Clutch Feedback Circuit High Voltage
P1546 - Air Conditioning [A/C] Clutch Feedback Circuit Low Voltage
P1571 - Traction Control Torque Request Circuit
P1574 - Stop lamp Switch Circuit
P1575 - Extended Travel Brake Switch Circuit High Voltage
P1626 - Theft Deterrent Fuel Enable Signal Not Received
P1630 - Theft Deterrent Learn Mode Active
P1631 - Theft Deterrent System - Password Incorrect
P1635 - 5 Volts Reference 1 Circuit

P1637 - Generator L-Terminal Circuit
P1638 - Generator F-Terminal Circuit
P1639 - 5 Volt Reference 2 Circuit
P1652 - Powertrain Induced Chassis Pitch Output Circuit
P1689 - Traction Control Delivered Torque Output Circuit
P1810 - TFP Valve Position Switch Circuit
P1860 - TCC PWM Solenoid Circuit - Electrical
P1868 - Transmission Fluid Life
P1870 - Transmission Component Slipping
P1875 - 4WD Low Switch Circuit
U1000 - Loss of Serial Communications for Class 2
U1026 - Lost Communications with Transmission Control System
U1040 - Loss of EBCM/EBTCM Communications
U1041 - Loss of EBCM Communication
U1056 - Loss of RTD Communication
U1064 - Loss of BCM/DIM/SBM Communications
U1096 - Loss of IPC Communications
U1153 - Loss of HVAC Communication
U1300 - Class 2 Communication Circuit High Voltage
Chevy Tracker/Sunrunner 2.0L
P1106 - Manifold Absolute Pressure [MAP] Sensor Circuit Intermittent High Voltage
P1107 - Manifold Absolute Pressure [MAP] Sensor Circuit Intermittent Low Voltage
P1111 - Intake Air Temperature [IAT] Sensor Circuit Intermittent High Voltage
P1112 - Intake Air Temperature [IAT] Sensor Circuit Intermittent Low Voltage
P1114 - Engine Coolant Temperature [ECT] Sensor Circuit Intermittent Low Voltage
P1115 - Engine Coolant Temperature [ECT] Sensor Circuit Intermittent High Voltage

P1121 - Throttle Position [TP] Sensor Circuit Intermittent High Voltage	
P1122 - Throttle Position [TP] Sensor Circuit Intermittent Low Voltage	
P1125 - Accelerator Pedal Position [APP] System	
P1133 - HO2S Insufficient Switching Bank 1 Sensor 1	
P1134 - HO2S Transition Time Ratio Bank1 Sensor 1	
P1153 - HO2S Insufficient Switching Bank 2 Sensor 1	
P1154 - HO2S Transition Time Ratio Bank 2 Sensor 1	
P1336 - Crankshaft Position [CKP] System Variation Not Learned	
P1345 - Crankshaft Position [CKP] Camshaft Position [CMP] Correlation	
P1351 - Ignition Control [IC] Circuit High Voltage	
P1361 - Ignition Coil Control Circuit Low Voltage	
P1380 - Misfire Detected - Rough Road Data Not Available	
P1381 - Misfire Detected - No Communication with Brake Control Module	
P1404 - Exhaust Gas Recirculation [EGR] Closed Position Performance	
P1415 - Secondary Air Injection [AIR] System Bank 1	
P1416 - Secondary Air Injection [AIR] System Bank 2	
P1441 - Evaporative Emission [EVAP] System Flow during Non-Purge	
P1442 - EVAP Monitor Circuit High Voltage during Ignition	
P1508 - Idle Speed Low - IAC System Not Responding	
P1509 - Idle Speed High - IAC System Not Responding	
P1621 - Control Module Long Term Memory Performance	
P1622 - Cylinder Select	
P1623 - Transmission Temperature Pull-Up Resistor	
P1625 - PCM System Reset	
P1626 - Theft Deterrent Fuel Enable Signal Not Received	
P1630 - Theft Deterrent Learn Mode Active	

P1631 - Theft Deterrent System - Password Incorrect

P1810 - TFP Valve Position SW. Circuit Malfunction

P1811 - Maximum Adapt and Long Shift

P1814 - Torque Converter Overstressed

P1860 - TCC PWM Solenoid Circuit - Electrical

P1870 - Transmission Component Slipping (See P1870 Diagnostic Test)

P1875 - 4WD Low Switch Circuit - Electrical

U1000 - Class 2 Communication Malfunction

U1026 - Loss of Class 2 Communication with ATC

U1041 - Loss of EBCM Communication

U1042 - Lost Communications with Brake/Traction Control System

U1064 - Loss of BCM and/or VTD Communication

U1192 - Loss of Passlock/EVO Class 2 Communications

U1193 - Loss of VIM Class 2 Communication

U1255 - Class 2 Communication Malfunction

U1301 - Class 2 Data Link High

Chevy Tracker/Sunrunner 1.6L

P1408 - Manifold Absolute Pressure [MAP] Sensor Circuit

P1410 - Fuel Tank Pressure Control Solenoid Control Circuit

P1450 - Barometric Pressure [BARO] Sensor Circuit

P1451 - Barometric Pressure [BARO] Sensor Performance

P1500 - Start Switch Circuit

P1510 - Control Module Long Term Memory Reset

P1530 - Ignition Timing Adjustment Switch Circuit

P1875 - 4WD Switch Circuit

Saturn and Cadillac Catera
P1106 - Manifold Absolute Pressure [MAP] Sensor Circuit Intermittent High Voltage
P1107 - Manifold Absolute Pressure [MAP] Sensor Circuit Intermittent Low Voltage
P1111 - Intake Air Temperature [IAT] Sensor Circuit Intermittent High Voltage
P1122 - Throttle Position [TP] Sensor Circuit Intermittent Low Voltage
P1125 - Accelerator Pedal Position [APP] System
P1133 - HO2S Insufficient Switching Bank 1 Sensor 1
P1134 - HO2S Transition Time Ratio Bank1 Sensor 1
P1137 - O2S Bank 1 Sensor 2 Lean System or Low Voltage
P1171 - Fuel System Lean during Acceleration
P1220 - Throttle Position [TP] Sensor 2 Circuit
P1221 - Throttle Position [TP] Sensor 1-2 Correlation
P1245 - Intake Plenum Switchover Valve
P1271 - Accelerator Pedal Position [APP] Sensor 1-2 Correlation
P1275 - Accelerator Pedal Position [APP] Sensor 1 Circuit
P1276 - Accelerator Pedal Position [APP] Sensor 1 Performance
P1277 - Accelerator Pedal Position Sensor 1 Circuit Low Voltage
P1278 - Accelerator Pedal Position Sensor 1 Circuit High Voltage
P1280 - Accelerator Pedal Position [APP] Sensor 2 Circuit
P1281 - Accelerator Pedal Position [APP] Sensor 2 Performance
P1282 - Accelerator Pedal Position Sensor 2 Circuit Low Voltage
P1283 - Accelerator Pedal Position Sensor 2 Circuit High Voltage
P1285 - Accelerator Pedal Position [APP] Sensor 3 Circuit
P1286 - Accelerator Pedal Position [APP] Sensor 3 Performance
P1336 - Crankshaft Position [CKP] System Variation Not Learned
P1351 - Ignition Control [IC] Circuit High Voltage

P1352 - Ignition Bypass Circuit High Voltage
P1361 - Ignition Control [IC] Circuit Low Voltage
P1362 - Ignition Bypass Circuit Low Voltage
P1374 - Crankshaft Position [CKP] High To Low Resolution Frequency Correlation
P1380 - Misfire Detected - Rough Road Data Not Available
P1381 - Misfire Detected - No Communication with Brake Control Module
P1403 - EGR Valve Position Sensor Circuit Low Voltage
P1404 - Exhaust Gas Recirculation [EGR] Closed Position Performance
P1405 - EGR Valve Position Sensor Circuit High Voltage
P1406 - Exhaust Gas Recirculation [EGR] Position Sensor Performance
P1441 - Evaporative Emission [EVAP] System Flow during Non-Purge
P1483 - Engine Cooling System Performance
P1510 - Throttle Control System - Limitation
P1511 - Throttle Control System - Limp Home System Performance
P1514 - Throttle Body Performance
P1515 - Commanded vs. Actual TP Correlation (TAC Module)
P1516 - Commanded vs. Actual Throttle Position Performance (TAC Module)
P1517 - Throttle Actuator Control [TAC] Module Performance
P1518 - PCM to TAC Module Serial Data Circuit
P1519 - Throttle Actuator Control (TAC) Module
P1520 - Park/Neutral Position [PNP] Switch Circuit
P1523 - Throttle Closed Position Performance
P1526 - TP Sensor Learn Not Done After Reprogramming
P1530 - Throttle Control System - Amplifier Adjustment
P1546 - Air Conditioning [A/C] Clutch Relay Control Circuit
P1554 - Cruise Control Feedback Circuit

P1566 - Engine RPM Too High - Cruise Control Disabled
P1571 - Traction Control Desired Torque Signal
P1573 - PCM/EBTCM or ABS/TCS Serial Data Circuit error
P1574 - Stop lamp Switch Circuit
P1575 - Extended Travel Brake Switch Circuit
P1585 - Cruise Lamp Control Circuit
P1599 - Engine Stall Or Near Stall Detected
P1601 - Serial Communication Problem with Device 1
P1602 - Knock Control Module Circuit
P1604 - Loss of IPC Serial Data
P1610 - Serial Communication Problem with Device 10
P1621 - Control Module Long Term Memory Performance
P1626 - Theft Deterrent Fuel Enable Signal Not Received
P1629 - Theft Deterrent Fuel Enable Signal Not Received
P1630 - Theft Deterrent System - ECM in Learn Mode
P1631 - Theft Deterrent System - Password Incorrect
P1632 - Theft Deterrent System - Fuel Disabled
P1635 - 5 Volt Reference 1 Circuit
P1637 - Charging System Lamp Control Circuit
P1639 - 5 Volt Reference 2 Circuit
P1640 - Driver 1
P1645 - Evaporative Emission [EVAP] Purge Solenoid Control Circuit
P1647 - Fuel Pump Relay Control Circuit
P1650 - Driver 2
P1653 - Low Engine Oil Level Lamp Control Circuit
P1655 - 1-2 Shift Solenoid Circuit

P1658 - Starter Relay Control Circuit

P1663 - Change Engine Oil Lamp Control Circuit

P1664 - Service Vehicle Soon Lamp Control Circuit

P1669 - ABS Unit Expected

P1673 - Engine Coolant Temperature Lamp Control Circuit

P1689 - Traction Control Delivered Torque Output Circuit

P1779 - Engine Torque Delivered to TCM Signal

P1780 - MIL Request by TCM

P1781 - Engine Torque Signal Circuit

P1791 - CAN Bus - Throttle Position

P1792 - CAN Bus - Engine Coolant

P1795 - CAN Bus - Throttle Body Position

P1810 - TFP Valve Position SW. Circuit Malfunction

P1811 - Maximum Adapt and Long Shift

P1812 - Transmission Over-temperature Condition

P1814 - Torque Converter Overstress

P1815 - Transmission Range Switch - Start In Wrong Range

P1816 - TFP Valve Position Switch - Park/Neu. With Drive Ratio

P1817 - TFP Valve Position Switch - Reverse With Drive Ratio

P1818 - TFP Valve Position Switch - Drive With Reverse Ratio

P1819 - Internal Mode Switch - No Start\Wrong Range

P1820 - Internal Mode Switch Circuit A - Low

P1822 - Internal Mode Switch Circuit B - High

P1823 - Internal Mode Switch Circuit P - Low

P1825 - Internal Mode Switch - Illegal Range

P1826 - Internal Mode Switch Circuit C - High

P1831 - PC Solenoid Power Circuit - Low Voltage
P1833 - A/T Solenoids Power Circuit - Low Voltage
P1842 - 1-2 Shift Solenoid Circuit - Low Voltage
P1843 - 1-2 Shift Solenoid Circuit - High Voltage
P1845 - 2-3 Shift Solenoid Circuit - Low Voltage
P1847 - 2-3 Shift Solenoid Circuit - High Voltage
P1860 - TCC PWM Solenoid Circuit - Electrical
P1864 - TCC Enable Solenoid Circuit
P1868 - Transmission Fluid Life
P1870 - Transmission Component Slipping
P1871 - Undefined Gear Ratio
P1886 - 3-2 Control Solenoid Electrical
P1887 - TCC Release Switch Circuit Malfunction
U1000 - No State of Health From Module
U1016 - Loss of PCM Communications
U1040 - Loss of EBCM/EBTCM Communications
U1041 - Loss of EBCM Communication
U1064 - Loss of Communication with BCM
U1088 - Loss of SDM Serial Data
U1096 - Loss of IPC Communications
U1152 - Loss of HVAC Serial Data
U1300 - Class 2 Shorted to Ground
U1301 - Class 2 Shorted to Battery +
U2100 - CAN Bus Communication
U2102 - More Controllers on Bus than Programmed
U2103 - Fewer Controllers on Bus than Programmed

U2104 - CAN Bus Reset Counter Overrun

U2105 - Lost Communications with Engine Control System

U2106 - Lost Communications with Transmission Control System

U2107 - Lost Communications with Body Control System

U2108 - Lost Communications with ABS/TCS Control System

All Other Vehicles

P1106 - Manifold Absolute Pressure [MAP] Sensor Circuit Intermittent High Voltage

P1107 - Manifold Absolute Pressure [MAP] Sensor Circuit Intermittent Low Voltage

P1111 - Intake Air Temperature [IAT] Sensor Circuit Intermittent High Voltage

P1112 - Intake Air Temperature [IAT] Sensor Circuit Intermittent Low Voltage

P1114 - Engine Coolant Temperature [ECT] Sensor Circuit Intermittent Low Voltage

P1115 - Engine Coolant Temperature [ECT] Sensor Circuit Intermittent High Voltage

P1120 - Throttle Position [TP] Sensor 1 Circuit

P1121 - Throttle Position [TP] Sensor Circuit Intermittent High Voltage

P1122 - Throttle Position [TP] Sensor Circuit Intermittent Low Voltage

P1125 - Accelerator Pedal Position [APP] System

P1133 - HO2S Insufficient Switching Bank 1 Sensor 1

P1134 - HO2S Transition Time Ratio Bank1 Sensor 1

P1137 - O2S Bank 1 Sensor 2 Lean System or Low Voltage

P1138 - O2S Bank 1 Sensor 2 Rich or High Voltage

P1153 - HO2S Insufficient Switching Bank 2 Sensor 1

P1154 - HO2S Transition Time Ratio Bank 2 Sensor 1

P1171 - Fuel System Lean during Acceleration

P1187 - Engine Oil Pressure Sensor Circuit Low Voltage

P1188 - Engine Oil Pressure Sensor Circuit High Voltage

P1191 - Intake Air Duct Leak

P1201 - (Alt. Fuel) Gas Mass Sensor Circuit Range/Performance
P1202 - (Alt. Fuel) Gas Mass Sensor Circuit Low Frequency
P1203 - (Alt. Fuel) Gas Mass Sensor Circuit High Frequency
P1214 - Injection Pump Timing Offset
P1215 - (Alt. Fuel) Quad Driver Output
P1217 - Fuel Solenoid Response Time Too Long
P1218 - Injection Pump Calibration Circuit
P1220 - Throttle Position [TP] Sensor 2 Circuit
P1221 - Throttle Position [TP] Sensor 1-2 Correlation
P1257 - Supercharge Engine Over boost
P1258 - Engine Coolant Over-temperature - Protection Mode Active
P1260 - Fuel Pump Speed Relay Control Circuit
P1270 - Accelerator Pedal Position Sensor A/D Converter Error
P1271 - Accelerator Pedal Position [APP] Sensor 1-2 Correlation
P1272 - Accelerator Pedal Position [APP] Sensor 2-3 Correlation
P1273 - Accelerator Pedal Position [APP] Sensor 1-3 Correlation
P1274 - Injector Wiring Incorrect
P1275 - Accelerator Pedal Position [APP] Sensor 1 Circuit
P1276 - Accelerator Pedal Position [APP] Sensor 1 Performance
P1277 - Accelerator Pedal Position [APP] Sensor 1 Circuit Low Voltage
P1278 - Accelerator Pedal Position [APP] Sensor 1 Circuit High Voltage
P1280 - Accelerator Pedal Position [APP] Sensor 2 Circuit
P1281 - Accelerator Pedal Position [APP] Sensor 2 Performance
P1282 - Accelerator Pedal Position [APP] Sensor 2 Circuit Low Voltage
P1283 - Accelerator Pedal Position [APP] Sensor 2 Circuit High Voltage
P1285 - Accelerator Pedal Position [APP] Sensor 3 Circuit

P1286 - Accelerator Pedal Position [APP] Sensor 3 Performance
P1287 - Accelerator Pedal Position [APP] Sensor 3 Circuit Low Voltage
P1288 - Accelerator Pedal Position [APP] Sensor 3 Circuit High Voltage
P1336 - Crankshaft Position [CKP] System Variation Not Learned
P1345 - Crankshaft Position [CKP] Camshaft Position [CMP] Correlation
P1345 - Crankshaft Position [CKP] Camshaft Position [CMP] Correlation
P1351 - Ignition Coil 1 Control Circuit High Voltage
P1352 - Ignition Coil 2 Control Circuit High Voltage
P1353 - Ignition Coil 3 Control Circuit High Voltage
P1354 - Ignition Coil 4 Control Circuit High Voltage
P1355 - Ignition Coil 5 Control Circuit High Voltage
P1356 - Ignition Coil 6 Control Circuit High Voltage
P1361 - Ignition Coil 1 Control Circuit Low Voltage
P1362 - Ignition Coil 2 Control Circuit Low Voltage
P1363 - Ignition Coil 3 Control Circuit Low Voltage
P1364 - Ignition Coil 4 Control Circuit Low Voltage
P1365 - Ignition Coil 5 Control Circuit Low Voltage
P1366 - Ignition Coil 6 Control Circuit Low Voltage
P1372 - Crankshaft Position [CKP] Sensor A-B Correlation
P1374 - 3X Reference Circuit
P1380 - Misfire Detected - Rough Road Data Not Available
P1381 - Misfire Detected - No Communication with Brake Control Module
P1404 - Exhaust Gas Recirculation [EGR] Closed Position Performance
P1405 - Exhaust Gas Recirculation System Valve 3
P1406 - Exhaust Gas Recirculation [EGR] Position Sensor Performance
P1407 - EGR Air Intrusion in Exhaust Supply to EGR Valve

P1408 - EGR Vacuum System Malfunction
P1409 - EGR Vacuum System Leak
P1410 - EGR Vacuum System Restriction
P1415 - Secondary Air Injection [AIR] System Bank 1
P1416 - Secondary Air Injection [AIR] System Bank 2
P1431 - Fuel Level Sensor 2 Performance
P1432 - Fuel Level Sensor 2 Circuit Low Voltage
P1433 - Fuel Level Sensor 2 Circuit High Voltage
P1441 - Evaporative Emission [EVAP] System Flow During Non-Purge
P1442 - EVAP Monitor Circuit High Voltage During Ignition On
P1483 - Engine Cooling System Performance
P1504 - Vehicle Speed Output Circuit
P1508 - Idle Speed Low - IAC System Not Responding
P1509 - Idle Speed High - IAC System Not Responding
P1514 - Throttle Body Performance
P1515 - Predicted vs. Actual Throttle Position Performance
P1516 - Commanded vs. Actual Throttle Position Performance (TAC Module)
P1517 - Throttle Actuator Control [TAC] Module Performance
P1518 - PCM To TAC Module Serial Data Circuit
P1519 - Throttle Actuator Control [TAC] Module Internal Circuit
P1520 - Park/Neutral Position [PNP] Switch Circuit
P1521 - P/N To D/R At High Throttle Angle - Power Reduction Mode Active
P1522 - Park/Neutral to Drive/Reverse at High RPM
P1523 - Throttle Closed Position Performance
P1539 - Air Conditioning [A/C] Clutch Feedback Circuit High Voltage
P1545 - Air Conditioning [A/C] Clutch Relay Control Circuit

P1546 - Air Conditioning [A/C] Clutch Feedback Circuit Low Voltage
P1548 - HVAC System Recirculation Door Motor Control Circuit
P1554 - Cruise Control Feedback Circuit
P1566 - Engine RPM Too High - Cruise Control Disabled
P1571 - Traction Control Torque Request Circuit
P1573 - PCM/EBTCM Serial Data Circuit
P1574 - Stop lamp Switch Circuit
P1575 - Extended Travel Brake Switch Circuit High Voltage
P1585 - Cruise Lamp Control Circuit
P1586 - Cruise Control Brake Switch 2 Circuit
P1599 - Engine Stall Or Near Stall Detected
P1601 - Serial Communication Problem with Device 1
P1602 - Serial Communication Problem with Device 2
P1604 - Loss of IPC Serial Data
P1605 - Loss of HVAC Serial Data
P1610 - Serial Communication Problem with Device 10
P1610 - Standard Body Module Serial Data Circuit
P1619 - Engine Oil Life Monitor Reset Switch Circuit
P1621 - EEPROM Write Error
P1622 - Cylinder Select
P1623 - Transmission Temperature Pull-Up Resistor
P1624 - PCM RAM Performance
P1625 - PCM System Reset
P1626 - Theft Deterrent Fuel Enable Signal Not Received
P1627 - Control Module Analog to Digital Performance
P1628 - Control Module Engine Coolant Temperature [ECT] Pull-up Resistor

P1629 - Anti-Theft Device Cranking Signal
P1630 - Theft Deterrent Learn Mode Active
P1631 - Theft Deterrent System - Password Incorrect
P1632 - Anti-Theft Device Fuel Disable
P1632 - Theft Deterrent System - Fuel Disabled
P1635 - 5 Volt Reference 1 Circuit
P1636 - PCM Stack Overrun
P1637 - Generator L-Terminal Circuit
P1638 - Generator F-Terminal Circuit
P1639 - 5 Volt Reference 2 Circuit
P1640 - Driver 1
P1641 - Malfunction Indicator Lamp [MIL] Control Circuit
P1643 - Wait to Start Lamp Control Circuit
P1644 - Traction Control Delivered Torque Output Circuit
P1646 - 5 Volt Reference 3 Circuit
P1650 - Driver 2
P1652 - Powertrain Induced Chassis Pitch Output Circuit
P1653 - Refer to Owner's Manual
P1654 - Reduced Engine Power Lamp Control Circuit
P1655 - EGR Solenoid Control Circuit
P1656 - Refer to Owner's Manual
P1658 - Starter Relay Control Circuit
P1662 - Cruise Lamp Control Circuit
P1663 - Driver 3 Line 3
P1664 - Service Vehicle Soon Lamp Control Circuit
P1673 - Engine Coolant Temperature Lamp Control Circuit

P1675 - Evaporative Emission [EVAP] Vent Solenoid Control Circuit
P1689 - Traction Control Delivered Torque Output Circuit
P1790 - Transmission Control Module Checksum
P1791 - Transmission Control Module Loop
P1792 - Transmission Control Module Reprogrammable Memory
P1793 - Transmission Control Module Stack Overrun
P1801 - Performance Selector Switch Failure
P1804 - Ground Control Relay
P1810 - Refer to Owner's Manual
P1811 - Maximum Adapt and Long Shift
P1812 - Transmission Over-temperature Condition
P1813 - Torque Control
P1814 - Torque Converter Overstressed
P1815 - Transmission Range Switch - Start In Wrong Range
P1816 - TFP Valve Position Sw. - Park/Neu. With Drive Ratio
P1817 - TFP Valve Position Sw. - Reverse With Drive Ratio
P1818 - TFP Valve Position Sw. - Drive Without Drive Ratio
P1819 - Internal Mode Switch - No Start\Wrong Range
P1820 - Internal Mode Switch Circuit A - Low
P1822 - Internal Mode Switch Circuit B - High
P1823 - Internal Mode Switch Circuit P - Low
P1825 - Internal Mode Switch - Illegal Range
P1826 - Internal Mode Switch Circuit C - High
P1835 - Kick-Down Switch
P1836 - Kick-Down Switch Failed Open
P1837 - Kick-Down Switch Failed Short

P1842 - 1-2 Shift Solenoid Circuit - Low Voltage
P1843 - 1-2 Shift Solenoid Circuit - High Voltage
P1845 - 2-3 Shift Solenoid Circuit - Low Voltage
P1847 - 2-3 Shift Solenoid Circuit - High Voltage
P1850 - Brake Band Apply Solenoid Circuit
P1851 - Brake Band Apply Solenoid Performance
P1852 - Brake Band Apply Solenoid Low Voltage
P1853 - Brake Band Apply Solenoid High Voltage
P1860 - Torque Converter Clutch PWM Solenoid Circuit
P1864 - TCC Enable Solenoid Circuit
P1868 - Transmission Fluid Life
P1870 - Transmission Component Slipping
P1871 - Undefined Gear Ratio
P1873 - TCC Stator Temperature Switch Circuit Low
P1874 - TCC Stator Temperature Switch Circuit High
P1875 - 4WD Low Switch Circuit - Electrical
P1886 - 3-2 Control Solenoid Electrical
P1887 - TCC Release Switch Circuit Malfunction
P1891 - Throttle Position Sensor PWM Signal Low
P1893 - Engine Torque Signal Low Voltage
P1894 - Engine Torque Signal High Voltage
P1895 - TCM to ECM Torque Reduction Circuit
U1000 - Expected Message not received
U1016 - Loss of PCM Communications
U1026 - Lost Communications with Transmission Control System
U1040 - Lost Communications with Brake/Traction Control System

U1041 - Lost Communications with Brake/Traction Control System
U1042 - Lost Communications with Brake/Traction Control System
U1056 - Loss of RTD Communication
U1064 - Lost communications with Body Multiple Control System
U1088 - Loss of SDM Serial Data
U1096 - Lost Communications with Driver Information & Displays Control System
U1152 - Lost communications with HVAC Control System
U1153 - Loss of HVAC Communication
U1192 - Loss of Passlock/EVO Class 2 Communications
U1193 - Loss of VIM and/or VTD Class 2 Communication
U1255 - Class 2 Communication Malfunction
U1300 - Class 2 Communication Circuit Short To Ground
U1301 - Class 2 Communication Circuit Short To Battery

With this chapter,
you will take your scan
tool to the next level
for Ford vehicles.
Even when you know
how to use the tool
to interrogate faults,
Ford, like every other manufacturer,
has its own ins and outs.

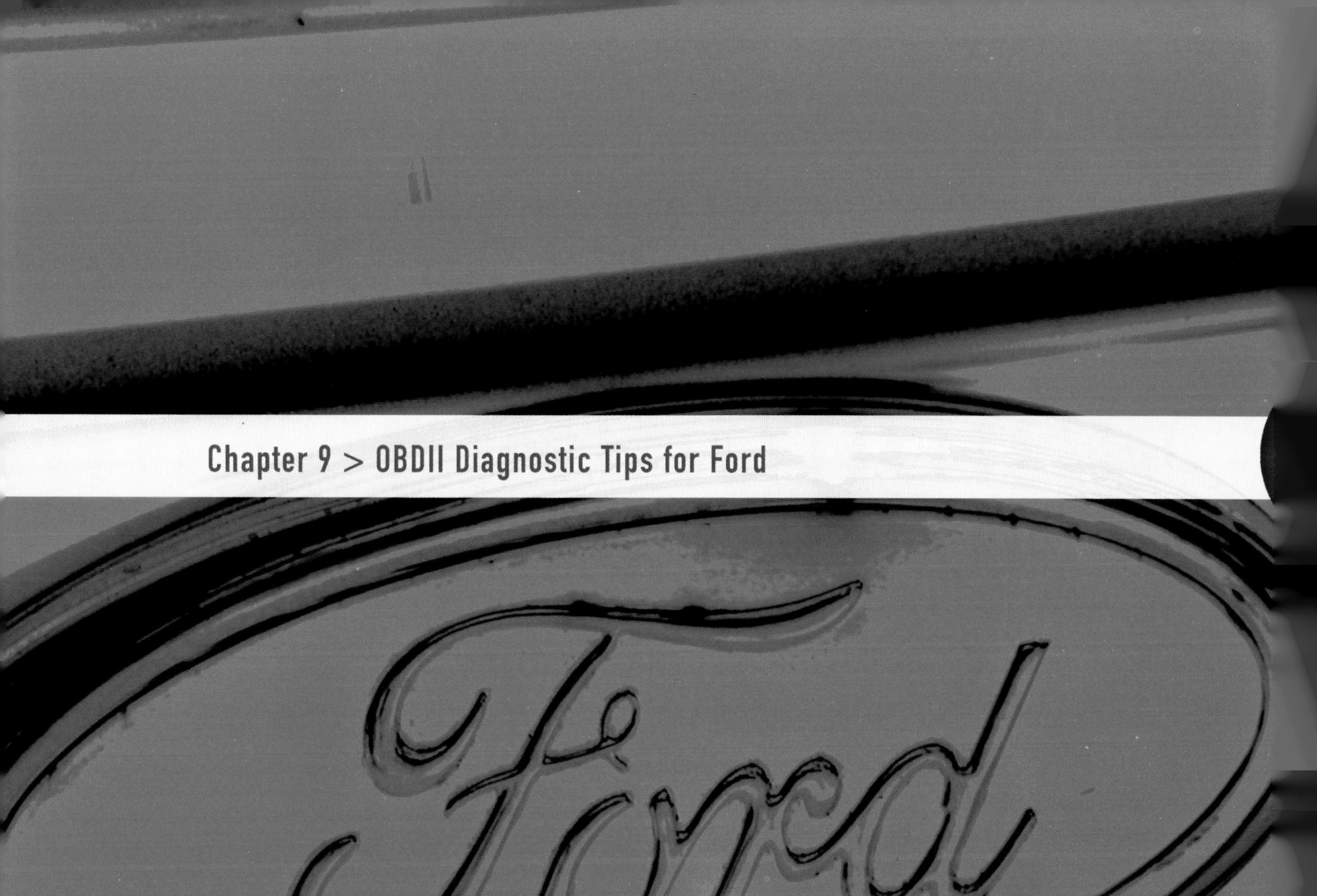

Chapter 9 > OBDII Diagnostic Tips for Ford

General Information about Ford

In 1994, Ford began their first adventure into OBDII with the Mustang (using the 3.8L V6 engines), Thunderbird and Cougar (using the 4.6L V8 engines). These models did not have everything needed to fully comply, but it was a start. They had the Misfire Monitor, operating at the OBDII threshold. They did not have the OBDII standard of the EVAP Monitor. Ford introduced OBDII with an EEC V control unit. In 1996 and beyond, Ford had to comply with the OBDII requirements. These requirements had to apply to both gas and diesel vehicles.

Understanding Ford Diagnostics

Ford, like GM, uses software called the Diagnostic Executive. This controls how the OBDII systems will be handled. The Diagnostic Executive software is found in the PCM. Here is a list of all the tasks of the Diagnostic Executive:

- Controls the Continuous and Non-Continuous Monitors
- Logs the Monitors and displays their status
- Controls the MIL to turn ON or OFF
- Records and clears the DTCs
- Records the Freeze Frame Data
- Controls non-OBDII related MILs

The main function of the Diagnostic Executive is to control the fault code categories, and handle the MIL when a fault has happened. The Diagnostic Executive will also control the turning off MIL. The Diagnostic Executive is the heart and soul of the OBDII system.

Ford's Driving Cycle

The purpose of the Drive Cycle is to run all the OBDII monitor tests. While on the road to test the Drive Cycle, connect a scan tool the to DLC. Have an assistant watch the Readiness status of the Non-Continuous and Continuous Monitors.

There are procedures that must be followed in order to clear the Ford P1000, I/M readiness code, if it is set. In addition, these procedures will prepare the Monitors to test their systems. The first 4 steps listed here are a prelude to the Monitor tests.

1. For the EVAP monitor to run, the ambient air temperature should be between 40 to 100°F and the altitude below 8000 feet. If the conditions are not at their specified values, the P1000 code cannot be cleared. The PCM must detect the stated conditions once (twice on some applications) before the EVAP monitor can be "bypassed" and the P1000 cleared.

2. The fuel tank level should be between 1/2 and 3/4's full. Preferably, the tank should be more than ¾ full. The Evaporative Monitor can only operate during the first 30 minutes of engine operation. When executing the procedure for this monitor, stay in part throttle mode and drive in a smooth fashion to minimize fuel slosh.

3. Install the scan tool. Turn the ignition key on, with the engine off. Cycle the key off, then on. Clear all DTCs.

4. Let the engine idle for 15 seconds, then drive at 40 MPH until the ECT is at least 170 ºF. Check if the IAT is within 40 to 100 ºF and the vehicle is not over 8000 ft. If it is not, the EVAP monitor will not complete (if not, complete the following tests, but please note that the last procedure will be required to "bypass" the EVAP monitor and clear the P1000 code).

Ford Specific Monitor Tests

After the above prelude to the driving cycle is completed, continue with the appropriate monitor below:

Catalyst Monitor

Purpose To monitor the catalyst's ability to accumulate and discharge oxygen needed to complete the emissions reducing chemical reactions. As the CAT ages, it's ability to store oxygen is lessened. This monitor checks the CATs ability to store and discharge oxygen.
Test Procedure Drive in stop-and-go traffic conditions. Include five different constant cruise speeds, ranging from 25 to 45 MPH over a 10-minute period.

EVAP Monitor

Purpose To test the fuel tank, and fuel delivery systems. The EVAP system prevents fuel tank vapors from entering the atmosphere. Fuel evaporation releases Hydrocarbons (HC). HC destroys the ozone.
This test will start the EVAP Monitor. I can't stress enough, if IAT is within 40 to 100°F, and the vehicle is not over 8000 ft, the test will run; if not, it will not run, and the PCM will not let you clear the P1000 code.
Test Procedure Let the vehicle cruise at 45 to 65 MPH for 10 minutes (avoid sharp turns and hills, so the fuel will not slosh and create more

vapor). To initiate the monitor, check the Real time data for:

- Throttle Position = 40 to 60 %
- Fuel Level Sensor = 15 to 85%

Heated Exhaust Gas Oxygen Sensor Monitor (or the HEGO)

Purpose This should start the HEGO monitor
Test Procedure Let the vehicle cruise at 40 MPH for up to 4 minutes.

EGR Monitor

Purpose This should start the EGR Monitor.

Test Procedure From a stop, accelerate to 45 MPH at 1/2 to 3/4 throttle. Repeat this 3 times.

Secondary Air Injection Monitor

Purpose This should start the Idle Speed Control portion of the CCM.

Test Procedure Bring the vehicle to a stop. Let the engine idle with transmission in drive (neutral for Manual Transmission) for 2 minutes.

Misfire & Fuel Monitors

Purpose This allows learning for the misfire monitor.
Test Procedure From a stop, accelerate the vehicle to 65 MPH. Then decelerate at closed throttle until 40 MPH (do not touch the throttle and brake pedals). Repeat this 3 times.

Pending Code Check and EVAP Monitor "Bypass" Check

Purpose This will determine if a pending code is preventing the clearing of P1000. This is important for Ford.

Test Procedure Check the scan tool, for pending codes. List the codes (if any), and repair. Then, re-run any incomplete monitor.
Note: If the EVAP monitor is not complete and IAT was out of the 40 to 100 °F temperature range, or the altitude is over 8000 ft., the EVAP "bypass" procedure must be followed:

Bypass Procedure Park the vehicle for a minimum of 8 hours. Repeat all Monitors listed above, but do not do the first 4 steps listed first.

Ford specific Fault codes (P1)

P1000 - OBD systems readiness test not complete
This fault is set when a fault is reset and one or more of the monitors have not been completed. Please refer to the drive cycle of this chapter.

P1001 - KOER not able to complete, KOER aborted
The Key On Engine Running test was not able to complete in the time that it was allowed to.

P1100 - Mass airflow sensor circuit intermittent
While the engine is running, the MAF sensor signal was out of range

P1101 - Mass airflow sensor out of self-test range
While the engine is running, the PCM detected the MAF sensor voltage was above or below its designated spec.

P1102 - Mass airflow sensor in range but lower than expected
While the engine is running, the PCM detected the MAF sensor voltage was below its designated spec.

P1103 - Mass air flow sensor in range but higher than expected
While the engine is running, the PCM detected the MAF sensor voltage was above its designated spec.

P1104 - Mass airflow sensor circuit ground
While the engine is running, the PCM detected the MAF sensor ground circuit was above or below its designated spec. Check wiring.

P1105 - Dual alternator upper fault
Check charging circuit.

P1106 - Dual alternator lower fault
Check charging circuit.

P1107 - Dual alternator lower circuit
Check charging circuit.

P1108 - Dual alternator battery lamp circuit
Check charging circuit MIL.

P1109 - Intake air temperature B circuit intermittent
While the engine is running, the PCM detected an intermittent open in circuit 2 of the IAT sensor. Check the Real Time data, and look for intermittent changes in the output.

P1110 - Intake air temperature B circuit open/short
While the engine is running, the PCM detected an open/short in circuit 2 of the IAT sensor. Check the Real Time data, and look for intermittent changes in the output.

P1111 - System pass
The PCM has detected that the system has passed its designated tests. In all honesty I checked this fault and have gotten no real answer.

P1112 - Intake air temperature circuit intermittent
While the engine is running, the PCM detected an intermittent open in circuit 1 of the IAT sensor. Check the Real Time data, and look for intermittent changes in the output.

P1113 - Intake air temperature circuit open/short
While the engine is running, the PCM detected an open/short in circuit 1 of the IAT sensor. Check the Real Time data, and look for intermittent changes in the output.

P1114 - Intake air temperature B circuit low input (super/turbo charged engines)
While the engine is running, the PCM detected the signal at the IAT sensor B to be less than .20 volts.

P1115 - Intake air temperature B circuit high input (super/turbo charged engines)
While the engine is running, the PCM detected the signal at the IAT sensor B to be greater than the maximum range.

P1116 - Engine coolant temperature sensor out of self-test range
While the engine is running, the PCM detected the ECT sensor voltage was above or below its designated spec of 0.3 to 3.7 volts.

P1117 - Engine coolant temperature sensor circuit intermittent
While the engine is running, the PCM detected an intermittent open in the ECT sensor. Check the Real Time data, and look for intermittent changes in the output.

P1118 - Manifold air temperature circuit low input
While the engine is running, the PCM detected the MAP sensor voltage was below its designated spec.

P1119 - Manifold air temperature circuit high input
While the engine is running, the PCM detected the MAP sensor voltage was above its designated spec.

P1120 - Throttle position sensor A out of range low
While the engine is running, the PCM detected the TPS voltage was below its designated spec.

P1121 - Throttle position sensor inconsistent with mass airflow sensor
While the engine is running, the PCM detected that the TPS and the MAF sensor are not consistent with their designated values. Check the Real Time data, TPS = < 4.82% @ a Load of 55% or TPS = > 49.05% @ a Load of 30%.

P1122 - Throttle position sensor A in range but lower than expected
While the engine is running, the PCM detected the TPS to be below its designated spec.

P1123 - Throttle position sensor A in range but higher than expected While the engine is running, the PCM detected the TPS to be above its designated spec.
P1124 - Throttle position sensor A out of self-test range While the engine is running, the PCM detected the TPS to be out its designated spec of less than 13.27% with KOEO, or more than 49.05% with KOER
P1125 - Throttle position sensor A intermittent While the engine is running, the PCM detected the TPS to be out its designated spec. Check the Real Time data, and do a wiggle test to look for intermittent changes in the output.
P1126 - Throttle position sensor circuit While the engine is running, the PCM detected the TPS to be out its designated spec
P1127 - Exhaust not warm, downstream O2 sensor not tested While the engine is running, the PCM detected the exhaust temp to be below its designated spec.
P1128 - Upstream HO2S sensors swapped You had better check how you installed the sensors, because the PCM has detected that they are not responding to the correct engine bank.
P1129 - Downstream HO2S sensors swapped You had better check how you installed the sensors, because the PCM has detected that they are not responding to the correct engine bank.
P1130 - Lack of HO2S11 switches - fuel trim at limit While the engine is running and in closed loop, the PCM detected the O2 sensors switching out of its designated spec, check for faults, or why the system's adaptation is at its limit
P1131 - Lack of HO2S11 switches - sensor indicates lean While the engine is running and in closed loop, the PCM detected the O2 sensors switching out of its designated spec, check for faults, or why the system's adaptation is at its lower limit
P1132 - Lack of HO2S11 switches - sensor indicates rich While the engine is running and in closed loop, the PCM detected the O2 sensors switching out of its designated spec, check for faults, or why the system's adaptation is at its upper limit
P1133 - Bank 1 fuel control shifted lean Check for rich condition on bank 1
P1134 - Bank 1 fuel control shifted rich Check for lean condition on bank 1

P1135 - Pedal position sensor A circuit intermittent While the engine is running, the PCM detected the pedal position sensor A to be out of its designated spec
P1135 - HO2S11 heater circuit low Check for O2 sensor heater, for bad connection, or relay bad, replace sensor if ok
P1136 - Control box fan circuit Check the cooling circuit of the PCM box.
P1136 - HO2S11 heater circuit high Check for O2 sensor heater, for short circuit, replace sensor if ok
P1137 - Lack of HO2S12 switches - sensor indicates lean While the engine is running and in closed loop, the PCM detected the O2 sensor switching out of its designated spec, check for faults, or why the system's adaptation is at its lower limit
P1138 - Lack of HO2S12 switches - sensor indicates rich While the engine is running and in closed loop, the PCM detected the O2 sensor switching out of its designated spec, check for faults, or why the system's adaptation is at its upper limit
P1139 - Water in fuel indicator circuit This fault is for diesels, the PCM detected that the water in fuel sensor circuit is defective; the indicator is in the fuel strainer
P1140 - Water in fuel condition This fault is for diesels, the PCM detected that there is water in the fuel.
P1141 - Fuel restriction indicator circuit This fault is for diesels, the PCM detected that the fuel restriction sensor circuit is defective; the sensor is in the fuel strainer
P1141 - HO2S12 heater circuit high Check for O2 sensor heater, for bad connection, or relay bad, replace sensor if ok
P1142 - Fuel restriction condition This fault is for diesels, the PCM detected that there is a restriction in the fuel delivery system, possible fuel filter clogged.
P1142 - HO2S12 heater circuit high Check for O2 sensor heater, for short circuit, replace sensor if ok
P1143 - Air assisted injector control valve range/performance While the engine is running, the PCM detected this control valve to be out its of designated spec

P1143 - Lack of HO2S switches, HO2S32 indicates lean
While the engine is running and in closed loop, the PCM detected the O2 sensor switching out of its designated spec, check for faults, or why the system's adaptation is at its lower limit

P1144 - Air assisted injector control valve circuit
While the engine is running, the PCM detected this control valve to be out its of designated spec

P1144 - Lack of HO2S switches, HO2S32 indicates rich
While the engine is running and in closed loop, the PCM detected the O2 sensor switching out of its designated spec, check for faults, or why the system's adaptation is at its upper limit

P1150 - Lack of HO2S21 switches - fuel trim at limit
While the engine is running and in closed loop, the PCM detected the O2 sensor switching out of its designated spec, check for faults, or why the system's adaptation is at its upper limit

P1151 - Lack of HO2S21 switches - sensor indicates lean
While the engine is running and in closed loop, the PCM detected the O2 sensor switching out of its designated spec, check for faults, or why the system's adaptation is at its lower limit

P1152 - Lack of HO2S21 switches - sensor indicates rich
While the engine is running and in closed loop, the PCM detected the O2 sensor switching out of its designated spec, check for faults, or why the system's adaptation is at its upper limit

P1153 - Bank 2 fuel control shifted lean
Check for rich condition on bank 2

P1154 - Bank 2 fuel control shifted rich
Check for rich condition on bank 2

P1156 - Fuel select switch circuit
While the engine is running, the PCM detected the fuel select switch to be out of its designated spec

P1157 - Lack of HO2S22 switches - sensor indicates lean
While the engine is running and in closed loop, the PCM detected the O2 sensor switching out of its designated spec, check for faults, or why the system's adaptation is at its lower limit

P1158 - Lack of HO2S22 switches - sensor indicates rich
While the engine is running and in closed loop, the PCM detected the O2 sensor switching out of its designated spec, check for faults, or why the system's adaptation is at its upper limit

P1159 - Fuel stepper motor
While the engine is running, the PCM detected the fuel stepper motor to be out of its designated spec

P1167 - Invalid test, operator did not actuate throttle While attempting to do an idle circuit test, the procedure was not followed.
P1168 - Fuel rail pressure sensor in range but low While the engine is running, the PCM detected the FRP sensor to be below its designated spec of 80 psi.
P1169 - Feedback A/F mixture control (HO2S12) While the engine is running, the PCM detected that the Feedback A/F mixture control to be out of its designated spec with the O2 sensor 1 - 2
P1170 - Feedback A/F mixture control (HO2S11) (For Gas engines) While the engine is running, the PCM detected that the Feedback A/F mixture control to be out of its designated spec with the O2 sensor 1 Ð 1
P1170 - Engine shut off solenoid (For diesel engines) While the engine is running, the PCM detected the fuel shut off to be out of its designated spec
P1171 - System too lean - banks 1 and 2 (lean fuel fault) (For Gas engines) While the engine is running and in closed loop, the PCM detected banks 1 and 2 to be out of its designated spec, check for faults, or why the system is at its lower limit
P1171 - Rotor sensor (For diesel engines) While the engine is running, the PCM detected the rotor sensor to be out of its designated spec
P1172 - System too rich - banks 1 and 2 (rich fuel fault) (For Gas engines) While the engine is running and in closed loop, the PCM detected banks 1 and 2 to be out of its designated spec, check for faults, or why the system is at its upper limit
P1172 - Rotor control (For diesel engines) While the engine is running, the PCM detected the rotor control to be out of its designated spec
P1173 - Feedback A/F mixture control (HO2S21) (For Gas engines) While the engine is running, the PCM detected the air/ fuel mixture to be out of its designated spec
P1173 - Rotor calibration (For diesel engines) While the engine is running, the PCM detected the rotor calibration to be out of its designated spec

P1174 - System too lean - banks 1 and 2 (suspect HO2S) While the engine is running and in closed loop, the PCM detected banks 1 and 2 to be out of its designated spec, check for faults, possible O2 sensor
P1175 - System too rich - banks 1 and 2 (suspect HO2S) While the engine is running and in closed loop, the PCM detected banks 1 and 2 to be out of its designated spec, check for faults, possible O2 sensor
P1176 - Long-term fuel trim too lean - banks 1 and 2 While the engine is running and in closed loop, the PCM detected banks 1 and 2 to be out of its designated spec, check for faults, or why the system's adaptation is at its lower limit
P1177 - Long-term fuel trim too rich - banks 1 and 2 While the engine is running and in closed loop, the PCM detected banks 1 and 2 to be out of its designated spec, check for faults, or why the system's adaptation is at its upper limit
P1178 - Long-term fuel trim too lean - banks 1 and 2 While the engine is running and in closed loop, the PCM detected banks 1 and 2 to be out of its designated spec, check for faults, or why the system's adaptation is at its lower limit
P1179 - Long-term fuel trim too rich - banks 1 and 2 While the engine is running and in closed loop, the PCM detected banks 1 and 2 to be out of its designated spec, check for faults, or why the system's adaptation is at its upper limit
P1180 - Fuel delivery system - low While the engine is running, the PCM detected the fuel delivery system to be below its designated spec.
P1181 - Fuel delivery system - high While the engine is running, the PCM detected the fuel delivery system to be above its designated spec.
P1182 - Fuel shut off solenoid circuit While the engine is running, the PCM detected the fuel shut off circuit to be out of its designated spec
P1183 - Engine oil temperature sensor circuit While the engine is running, the PCM detected the EOT to be out of its designated spec; possibly sensor was shorted to power or open circuit.
P1184 - Engine oil temperature sensor out of self-test range While the engine is running, the PCM detected the EOT to be out of its designated spec of 0.3 to 1.2 volts
P1185 - O2 sensor heater circuit open - hardware fault (For Gas engines) While the engine is running, the PCM detected the O2 sensor to be out of its designated spec; possible bad O2 sensor, or faulty wiring

P1185 - Fuel pump temperature sensor high (For Diesel engines) While the engine is running, the PCM detected the Fuel pump temp sensor to be out of its designated spec
P1186 - O2 sensor heater circuit shorted - hardware fault **P1186** - Fuel pump temperature sensor low (For Diesel engines) While the engine is running, the PCM detected the Fuel pump temp sensor to be out of its designated spec
P1187 - O2 sensor heater circuit open While the engine is running, the PCM detected the O2 sensor circuit was above or below its designated spec. Check wiring.
P1188 - O2 sensor heater circuit resistance While the engine is running, the PCM detected the O2 sensor circuit was above or below its designated spec. Check wiring.
P1189 - O2 sensor heater circuit low resistance fault 1 While the engine is running, the PCM detected the O2 sensor circuit was above or below its designated spec. Check wiring.
P1189 - Pump speed signal (For Diesel engines) While the engine is running, the PCM detected the Fuel pump speed sensor to be out of its designated spec
P1190 - O2 sensor heater circuit low resistance fault 2 While the engine is running, the PCM detected the O2 sensor circuit was above or below its designated spec. Check wiring.
P1191 - O2 sensor heater circuit open - hardware fault While the engine is running, the PCM detected the O2 sensor circuit was above or below its designated spec. Check wiring.
P1192 - O2 sensor heater circuit shorted While the engine is running, the PCM detected the O2 sensor circuit was above or below its designated spec. Check wiring.
P1193 - O2 sensor heater circuit open - inferred fault While the engine is running, the PCM detected the O2 sensor circuit was above or below its designated spec. Check wiring.

P1194 - O2 sensor heater circuit resistance fault
While the engine is running, the PCM detected the O2 sensor circuit was above or below its designated spec. Check wiring.

P1195 - O2 sensor heater circuit low resistance fault 1
While the engine is running, the PCM detected the O2 sensor circuit was above or below its designated spec. Check wiring.

P1195 - BARO sensor circuit
While the engine is running, the PCM detected the BARO sensor to be out of its designated spec

P1196 - O2 sensor heater circuit low resistance fault 2

P1196 - Key off voltage high
While the engine is running, the PCM detected the key off voltage to be above its designated spec.

Check ignition switch

P1197 - Mileage switch circuit
While the engine is running, the PCM detected the mileage switch to be out of its designated spec

P1198 - Fuel level input circuit high
While the engine is running, the PCM detected the Fuel level input to be above its designated spec.

P1199 - Fuel level input circuit low
While the engine is running, the PCM detected the fuel level input to be below its designated spec.

P1201 - Cylinder #1 injector circuit open/shorted
While the engine is running, the PCM detected the injector circuit was above or below its designated spec. Check wiring.

P1202 - Cylinder #2 injector circuit open/shorted
While the engine is running, the PCM detected the injector circuit was above or below its designated spec. Check wiring.

P1203 - Cylinder #3 injector circuit open/shorted
While the engine is running, the PCM detected the injector circuit was above or below its designated spec. Check wiring.

P1204 - Cylinder #4 injector circuit open/shorted
While the engine is running, the PCM detected the injector circuit was above or below its designated spec. Check wiring.

P1205 - Cylinder #5 injector circuit open/shorted While the engine is running, the PCM detected the injector circuit was above or below its designated spec. Check wiring.
P1206 - Cylinder #6 injector circuit open/shorted While the engine is running, the PCM detected the injector circuit was above or below its designated spec. Check wiring.
P1207 - Cylinder #7 injector circuit open/shorted While the engine is running, the PCM detected the injector circuit was above or below its designated spec. Check wiring.
P1208 - Cylinder #8 injector circuit open/shorted While the engine is running, the PCM detected the injector circuit was above or below its designated spec. Check wiring.
P1209 - Injector control pressure peak delta test fault While the engine is running, the PCM detected the injector circuit was above or below its designated spec. Pressure check the system.
P1210 - Injector control pressure above expected level While the engine is running, the PCM detected the injector circuit was above or below its designated spec. Pressure check the system
P1211 - Injector control pressure above/below desired While the engine is running, the PCM detected the injector circuit was above or below its designated spec. Pressure check the system.
P1212 - Injector control pressure not at expected level While the engine is running, the PCM detected the injector circuit was above or below its designated spec. Pressure check the system.
P1213 - Start injector circuit While the engine is running, the PCM detected the injector circuit was above or below its designated spec. Check the cold start system.
P1214 - Pedal position sensor B circuit intermittent While the engine is running, the PCM detected the PPS B circuit was above or below its designated spec. Check the Real Time data, and look for intermittent changes in the output.

P1215 - Pedal position sensor C circuit low input While the engine is running, the PCM detected the PPS C circuit was below its designated spec. Check the Real Time data, and look for intermittent changes in the output.
P1216 - Pedal position sensor C circuit high input While the engine is running, the PCM detected the PPS C circuit was above its designated spec. Check the Real Time data, and look for intermittent changes in the output.
P1217 - Pedal position sensor C circuit intermittent While the engine is running, the PCM detected the PPS C circuit was above or below its designated spec. Check the Real Time data, and look for intermittent changes in the output.
P1220 - Series throttle control system While the engine is running, the PCM detected a fault in the STC system.
P1221 - Traction control system While the engine is running, the PCM detected a fault in the traction control system.
P1222 - Traction control output circuit While the engine is running, the PCM detected a fault in the traction control system.
P1223 - Pedal position sensor B circuit high input While the engine is running, the PCM detected the TPS B to be above its designated spec.
P1224 - Throttle position sensor B out of self-test range While the engine is running, the PCM detected the TPS B to be out of its designated spec
P1225 - Needle lift sensor While the engine is running, the PCM detected the NLS to be out of its designated spec
P1227 - Timing governor While the engine is running, the PCM detected the Timing governor to be out of its designated spec
P1228 - Wastegate failed open (under pressure) While the engine is running, the PCM detected that the Wastegate was stuck, or damaged. Check for carbon build up at Wastegate.
P1229 - Charge air cooler pump driver While the engine is running, the PCM detected the air cooler pump driver to be out of its designated spec. Check the intercooler pump circuit.
P1230 - Fuel pump low speed malfunction While the engine is running, the PCM detected excessive current or voltage to the fuel pump.

P1231 - Fuel pump secondary circuit low, high speed While the engine is running, the PCM detected low current or voltage to the secondary fuel pump.
P1232 - Fuel pump speed primary circuit (two speed fuel pump) While the engine is running, the PCM detected excessive current or voltage to the fuel pump.
P1233 - Fuel pump driver module disabled or off line While the engine is running, the PCM detected a low signal to the fuel pump.
P1234 - Fuel pump driver module disabled or off line (fuel pump driver module) While the engine is running, the PCM detected a low signal to the fuel pump.
P1235 - Fuel pump control out of range (fuel pump driver module) While the engine is running, the PCM detected a low signal or no signal to the fuel pump.
P1236 - Fuel pump control out of range (fuel pump driver module) While the engine is running, the PCM detected a low signal to the fuel pump.
P1237 - Fuel pump secondary circuit (fuel pump driver module) While the engine is running, the PCM detected a fault in the secondary fuel pump system.
P1238 - Fuel pump secondary circuit (fuel pump driver module) While the engine is running, the PCM detected a fault in the secondary fuel pump system.
P1239 - Speed fuel pump positive feed While the engine is running, the PCM detected excessive current or voltage to the fuel pump.
P1240 - Sensor power supply While the engine is running, the PCM detected a problem with the power supply to the fuel pump.
P1241 - Sensor power supply low input While the engine is running, the PCM detected a problem with the power supply to the fuel pump.
P1242 - Sensor power supply high input While the engine is running, the PCM detected a problem with the power supply to the fuel pump.
P1243 - Second fuel pump fault of ground fault While the engine is running, the PCM detected a problem with the ground circuit to the fuel pump.
P1244 - Alternator load high input While the engine is running, the PCM detected the GLS to be above its designated spec.
P1245 - Alternator load low input While the engine is running, the PCM detected the GLS to be below its designated spec.

P1246 - Alternator load input While the engine is running, the PCM detected the GLS to be above or below its designated spec.
P1247 - Turbo boost pressure low While the engine is running, the PCM detected the boost pressure to be below its designated spec. Check boost pressure.
P1248 - Turbo boost pressure not detected While the engine is running, the PCM detected the boost pressure to be out of its designated spec. Check boost pressure.
P1249 - Wastegate control valve performance While the engine is running, the PCM detected that the Wastegate was stuck, or damaged.
P1250 - Fuel pressure regulator control solenoid While the engine is running, the PCM detected an open/short in the fuel pressure regulator control solenoid.
P1251 - Air mixture solenoid circuit While the engine is running, the PCM detected the air mixture solenoid to be out of its designated spec
P1259 - Immobilizer to PCM signal error While the engine is running, the PCM detected that the Immobilizer circuit to be out of its designated spec. Check if PCM was improperly flashed, or the wrong key was used.
P1260 - Theft detected, vehicle immobilized While the engine is running, the PCM detected that a theft has occurred.
P1261 - Cylinder #1 high to low side short While the engine is running, the PCM detected that a short has occurred.
P1262 - Cylinder #2 high to low side short While the engine is running, the PCM detected that a short has occurred.
P1263 - Cylinder #3 high to low side short While the engine is running, the PCM detected that a short has occurred.
P1264 - Cylinder #4 high to low side short While the engine is running, the PCM detected that a short has occurred.
P1265 - Cylinder #5 high to low side short While the engine is running, the PCM detected that a short has occurred.
P1266 - Cylinder #6 high to low side short While the engine is running, the PCM detected that a short has occurred.

P1267 - Cylinder #7 high to low side short While the engine is running, the PCM detected that a short has occurred.
P1268 - Cylinder #8 high to low side short While the engine is running, the PCM detected that a short has occurred.
P1269 - Immobilizer code not programmed While the engine is running, the PCM detected that the Immobilizer circuit to be out of its designated spec. Check if PCM was improperly flashed, or the wrong key was used.
P1270 - Engine RPM or vehicle speed limiter reached While the engine is running, the PCM detected that the vehicle has exceeded its maximum limit. If engine is running poorly, check for bent valves.
P1271 - Cylinder #1 high to low side open While the engine is running, the PCM detected an open circuit has occurred.
P1272 - Cylinder #2 high to low side open While the engine is running, the PCM detected an open circuit has occurred.
P1273 - Cylinder #3 high to low side open While the engine is running, the PCM detected an open circuit has occurred.
P1274 - Cylinder #4 high to low side open While the engine is running, the PCM detected an open circuit has occurred.
P1275 - Cylinder #5 high to low side open While the engine is running, the PCM detected an open circuit has occurred.
P1276 - Cylinder #6 high to low side open While the engine is running, the PCM detected an open circuit has occurred.
P1277 - Cylinder #7 high to low side open While the engine is running, the PCM detected an open circuit has occurred.
P1278 - Cylinder #8 high to low side open While the engine is running, the PCM detected an open circuit has occurred.
P1285 - Cylinder head over temperature condition While the engine is running, the PCM detected an overheat condition at the cylinder head.
P1286 - Fuel pulse width in range but lower than expected While the engine is running, the PCM detected the fuel pulse width to be switching below its designated spec.

P1287 - Fuel pulse width in range but higher than expected While the engine is running, the PCM detected the fuel pulse width to be switching above its designated spec.
P1288 - Cylinder head temperature sensor out of self test range While the engine is running, the PCM detected that the sensor is not operating at its desired range.
P1289 - Cylinder head temperature sensor circuit high input While the engine is running, the PCM detected a short circuit in the cyl head temp sensor
P1290 - Cylinder head temperature sensor circuit low input While the engine is running, the PCM detected an open circuit in the cyl head temp sensor
P1299 - Cylinder head over temperature protection active While the engine is running, the PCM detected an overheat condition. The cooling system will go into a fail safe mode when this condition is detected.
P1300 - Boost calibration fault While the engine is running, the PCM detected fault in the boost pressure.
P1301 - Boost calibration high While the engine is running, the PCM detected the boost pressure to be out of its designated spec. Check boost pressure.
P1302 - Boost calibration low While the engine is running, the PCM detected the boost pressure to be out of its designated spec. Check boost pressure.
P1303 - Exhaust gas recirculation calibration fault While the engine is running, the PCM detected fault in the boost pressure.
P1304 - Exhaust gas recirculation calibration high While the engine is running, the PCM detected the EGR circuit to be out of its designated spec.
P1305 - Exhaust gas recirculation calibration low While the engine is running, the PCM detected the EGR circuit to be out of its designated spec.
P1306 - Kickdown relay pull-in circuit While the engine is running, the PCM detected fault in the Kickdown relay circuit.
P1307 - Kickdown relay hold circuit While the engine is running, the PCM detected fault in the Kickdown relay circuit.
P1308 - A/C clutch circuit While the engine is running, the PCM detected fault in the A/C clutch circuit.

P1309 - Misfire monitor AICE chip fault While the engine is running, the PCM disabled the misfire monitor. Check the Real Time data, and look for the Cam Position sensor output to be low.
P1312 - Injection pump timing actuator circuit While the engine is running, the PCM detected the injection pump timing to be out of its designated spec
P1313 - Misfire rate catalyst damage fault - bank 1 While the engine is running, the PCM detected the misfire rate to be out of its designated spec
P1314 - Misfire rate catalyst damage fault - bank 2 While the engine is running, the PCM detected the misfire rate to be out of its designated spec
P1315 - Persistent misfire While the engine is running, the PCM detected the misfire rate to be out of its designated spec
P1318 - Injection timing piston position sensor circuit While the engine is running, the PCM detected that the injection timing piston position sensor to be faulty. Check the Real Time data, and look for intermittent changes in the output.
P1319 - Injection timing piston position sensor circuit range/performance While the engine is running, the PCM detected that the injection timing piston position sensor to be out of its designated spec
P1336 - Crankshaft/camshaft sensor range/performance While the engine is running, the PCM detected the crank/cam sensor voltage was above or below its designated spec.
P1340 - Camshaft position sensor B circuit While the engine is running, the PCM detected the CMP sensor B to be out of its designated spec
P1341 - Camshaft position sensor B circuit range/performance While the engine is running, the PCM detected the CMP sensor B voltage was above or below its designated spec.
P1345 - Cylinder discrimination signal (from CMP sensor) While the engine is running, the PCM detected the CMP sensor B to be out of its designated spec
P1346 - Fuel level sensor B circuit While the engine is running, the PCM detected the fuel level sensor B to be out of its designated spec
P1347 - Fuel level sensor B circuit range/performance While the engine is running, the PCM detected the fuel level sensor B voltage was above or below its designated spec.

P1348 - Fuel level sensor B circuit low input
While the engine is running, the PCM detected the fuel level sensor B to be below its designated spec.

P1349 - Fuel level sensor B circuit high input
While the engine is running, the PCM detected the fuel level sensor B to be above its designated spec.

P1350 - Fuel level sensor B circuit intermittent
While the engine is running, the PCM detected that the fuel level sensor B to be out of its designated spec

P1351* - Ignition diagnostic monitor input circuit

P1352* - Ignition coil A primary circuit

P1353* - Ignition coil B primary circuit

P1354* - Ignition coil C primary circuit

P1355* - Ignition coil D primary circuit

P1356* - Ignition diagnostic monitor indicates engine not turning

P1357* - Ignition diagnostic monitor pulse width not defined

P1358* - Ignition diagnostic monitor signal out of self-test range (no CPU OK)

P1359* - Spark output circuit

P1360* - Ignition coil A secondary circuit

P1361* - Ignition coil B secondary circuit

P1362* - Ignition coil C secondary circuit

P1363* - Ignition coil D secondary circuit

P1364* - Ignition coil primary circuit

P1365* - Ignition coil secondary circuit

P1369* - Engine temperature light circuit

P1370* - Insufficient RPM increase during spark test

P1371* - Ignition coil - cylinder 1 - early activation fault

P1372* - Ignition coil - cylinder 2 - early activation fault

P1373* - Ignition coil - cylinder 3 - early activation fault

P1374* - Ignition coil - cylinder 4 - early activation fault

P1375* - Ignition coil - cylinder 5 - early activation fault

***Note:** faults P1351 to 1376 are related to the ignition system; check the system, since it is out of its designated spec.

P1376* - Ignition coil - cylinder 6 - early activation fault
P1380 - Camshaft position actuator circuit (bank 1) While the engine is running, the PCM detected a short/open circuit to the variable cam actuator bank 1
P1381 - Camshaft position timing over advanced (bank 1) While the engine is running, the PCM detected the variable cam actuator to be above its designated spec.
P1382 - Camshaft position timing solenoid #1 circuit While the engine is running, the PCM detected a short/open circuit to the variable cam solenoid #1
P1383 - Camshaft position timing over retarded (bank 1) While the engine is running, the PCM detected the cam timing to be below its designated spec.
P1384 - Variable valve timing solenoid A circuit While the engine is running, the PCM detected a short/open circuit to the variable timing solenoid A.
P1385 - Camshaft position actuator circuit (bank 2) While the engine is running, the PCM detected a short/open circuit to the variable cam actuator bank 2
P1386 - Camshaft position timing over advanced (bank 2) While the engine is running, the PCM detected the variable cam actuator to be above its designated spec.
P1387 - Camshaft position timing solenoid #2 circuit While the engine is running, the PCM detected a short/open circuit to the variable cam solenoid #2
P1388 - Camshaft position timing over retarded (bank 2) While the engine is running, the PCM detected the cam timing to be below its designated spec.
P1389 - Glow plug circuit high side, low input While the engine is running, the PCM detected the glow plug circuit to be below its designated spec. Check the power to the glow plug circuit.
P1390 - Octane adjust service pin in use/circuit open Better check to see if you removed the service pin. For the PCM is detecting that it is in place.
P1391 - Glow plug circuit low input (bank 1) While the engine is running, the PCM detected the glow circuit bank 1 to be below its designated spec.
P1392 - Glow plug circuit high input (bank 1) While the engine is running, the PCM detected the glow circuit bank 1 to be above its designated spec.
P1393 - Glow plug circuit low input (bank 2) While the engine is running, the PCM detected the glow circuit bank 2 to be below its designated spec.

***Note:** faults P1351 to 1376 are related to the ignition system; check the system, since it is out of its designated spec.

P1394 - Glow plug circuit high input (bank 2)
While the engine is running, the PCM detected the glow circuit bank 2 to be above its designated spec.

P1395 - Glow plug monitor fault (bank 1)
While the engine is running, the PCM detected that there is a fault in the glow plug circuit bank 1.

P1396 - Glow plug monitor fault (bank 2)
While the engine is running, the PCM detected that there is a fault in the glow plug circuit bank 2.

P1397 - System voltage out of self-test range
While the engine is running, the PCM detected the system voltage was above or below its designated spec of 12.1 to 14.5 volts.

P1398 - Variable valve timing solenoid B circuit high input
While the engine is running, the PCM detected the variable valve timing solenoid B to be above its designated spec.

P1399 - Glow plug circuit high side, high input
While the engine is running, the PCM detected the glow plug circuit to be above its designated spec.

P1400 - Differential pressure feedback EGR circuit low input
While the engine is running, the PCM detected the DPF EGR sensor to be below its designated spec of 0.2 volts.

P1401 - Differential pressure feedback EGR circuit high input
While the engine is running, the PCM detected the DPF EGR sensor to be above its designated spec of 4.5 volts

P1402 - Exhaust gas recirculation valve position sensor circuit
While the engine is running, the PCM detected a short/open circuit to the EGR valve position sensor.

P1403 - Differential pressure feedback sensor hoses reversed
I think you know the drill, check the hoses; you may have put them on backwards, because the PCM has detected a fault.

P1404 - EGR temperature sensor circuit
While the engine is running, the PCM detected a short/open circuit to the EGR temp sensor

P1405 - Differential pressure feedback sensor upstream hose off or plugged
I think this one speaks for itself.

P1406 - Differential pressure feedback sensor downstream hose off of plugged
Again, I think this one speaks for itself.

P1407 - Exhaust gas recirculation no flow detected And one more time, I think this one speaks for itself. Still check to see if the ERG valve is stuck, or carboned up.
P1408 - Exhaust gas recirculation flow out of self test range While the engine is running, the PCM detected the ERG flow to be out of its designated spec for the lack of movement during its test.
P1409 - EGR vacuum regulator solenoid circuit While the engine is running, the PCM detected the EGR vacuum regulator solenoid voltage was above or below its designated spec.
P1411 - Secondary air injection incorrect downstream flow detected While the engine is running, the PCM detected the amount of air to be below its designated spec at the O2 sensor, during the operation of the secondary air injection.
P1413 - Secondary air injection monitor circuit low input While the engine is running, the PCM detected the secondary air monitor to be below its designated spec.
P1414 - Secondary air injection monitor circuit high input While the engine is running, the PCM detected the secondary air monitor to be above its designated spec.
P1415 - Air pump circuit While the engine is running, the PCM detected that there is a fault in the air pump circuit
P1420 - Catalyst temperature sensor While the engine is running, the PCM detected that there is a fault in the cat temp sensor. Check the temp of the cat, and determine if the cat is plugged, or sensor is bad.
P1421 - Catalyst damage While the engine is running, the PCM detected that there is a fault in the cat. This can happen if the engine has been running too rich for a prolonged period of time, or if the engine has been burning oil.
P1422 - Exhaust gas ignition temperature sensor While the engine is running, the PCM detected the exhaust gas ignition temperature sensor to be out of its designated spec.
P1423 - Exhaust gas ignition functional test While the engine is running, the PCM detected the exhaust gas ignition temperature sensor voltage was above or below its designated spec.
P1429 - Electric air pump primary While the engine is running, the PCM detected the primary air pump to be out of its designated spec

P1430 - Electric air pump secondary
While the engine is running, the PCM detected the secondary air pump to be out of its designated spec

P1432 - Thermostat heater control circuit
While the engine is running, the PCM detected the thermostat heater control voltage was above or below its designated spec.

P1433 - A/C refrigerant temperature circuit low
While the engine is running, the PCM detected the A/C temp circuit to be below its designated spec.

P1434 - A/C refrigerant temperature circuit high
While the engine is running, the PCM detected the A/C temp circuit to be above its designated spec.

P1435 - A/C refrigerant temperature circuit range/performance
While the engine is running, the PCM detected the A/C temp circuit to be out of its designated spec

P1436 - A/C evaporator air temperature circuit low
While the engine is running, the PCM detected the evaporator temp circuit to be below its designated spec.

P1437 - A/C evaporator air temperature circuit high
While the engine is running, the PCM detected the evaporator temp circuit to be above its designated spec.

P1438 - A/C evaporator air temperature circuit range/performance
While the engine is running, the PCM detected the evaporator temp circuit to be out of its designated spec

P1440 - Purge valve stuck open

P1441 - ELC system 1

P1442 - Evaporative emission control system control leak detected
During cold start, the PCM detected a small leak in the EVAP system. Leak = < 0.040”

P1443 - Evaporative emission control system control valve
During cold start, the PCM detected a leak or blockage between the canister purge valve, intake manifold and the EVAP canister.

P1444 - Purge flow sensor circuit low input
While the engine is running, the PCM detected the PF sensor to be below its designated spec.

P1445 - Purge flow sensor circuit high input
While the engine is running, the PCM detected the PF sensor to be above its designated spec.

P1446 - Evaporative vacuum solenoid circuit
While the engine is running, the PCM detected the evaporative vacuum solenoid to be out of its designated spec

P1449 - Evaporative check solenoid circuit While the engine is running, during the warm up stage and the vehicle at a steady speed, the PCM detected the EVAP system not to be holding vacuum.
P1450 - Unable to bleed up fuel tank vacuum While the engine is running, during the warm up stage and the vehicle at a steady speed, the PCM detected the fuel tank vacuum at above idle speed to be above its designated spec.
P1451 - Evaporative emission control system vent control circuit While the engine is running, the canister vent solenoid enabled and the vehicle at a steady speed, the PCM detected that the solenoid voltage was above or below its designated
P1452 - Unable to bleed up fuel tank vacuum While the engine is running, during the warm up stage and the vehicle at a steady speed, the PCM detected the fuel tank vacuum at above idle speed to be above its designated spec.
P1453 - Fuel tank pressure relief valve malfunction While the engine is running, during the warm up stage and the vehicle at a steady speed, the PCM detected that the fuel tank pressure relief valve is not holding vacuum
P1454 - Evaporative emission control system vacuum test While the engine is running, during the warm up stage and the vehicle at a steady speed, the PCM detected that the system vacuum test could not be completed
P1455 - Evaporative emission control system control leak detected (gross leak/no flow) Gas Cap Gas Cap Gas Cap Gas Cap Gas Cap Gas Cap Gas Cap Gas Cap Gas Cap, did you get what I am talking about, check the Gas Cap
P1456 - Fuel tank temperature sensor circuit While the engine is running, the PCM detected the tank temp sensor voltage was above or below its designated spec.
P1457 - Purge solenoid control system While the engine is running, the PCM detected that there is a fault in the purge solenoid control system
P1457 - Unable to pull fuel tank vacuum While the engine is running, during the warm up stage and the vehicle at a steady speed, the PCM detected that vacuum could not be drawn for the fuel tank
P1460 - Wide open throttle A/C cutout circuit While the engine is running, the PCM detected that there is a fault in the WOT A/C cutout. Check if the A/C WOT relay is operation properly. Check if proper voltage is applied during WOT.

P1461 - A/C pressure sensor circuit high input While the engine is running, the PCM detected the A/C pressure sensor to be above its designated spec.
P1462 - A/C pressure sensor circuit low input While the engine is running, the PCM detected the A/C pressure sensor to be below its designated spec.
P1463 - A/C pressure sensor insufficient pressure change Check for a leak in the A/C system. Easy one huh. Just checking to see if you are still paying attention.
P1464 - A/C demand out of self test range While the engine is running and during its self test, the PCM detected the A/C demand signal to be above its designated spec.
P1465 - A/C relay circuit While the engine is running, the PCM detected the A/C relay voltage was above or below its designated spec.
P1466 - A/C refrigerant temperature sensor circuit While the engine is running, the PCM detected the refrigerant temp sensor voltage was above or below its designated spec.
P1467 - A/C compressor temperature sensor While the engine is running, the PCM detected the compressor speed sensor voltage was above or below its designated spec.
P1469 - Rapid A/C cycling While the engine is running, the PCM detected the A/C compressor to be cycling too many times. This usually means that the refrigerant level is too low.
P1470 - A/C cycling period too short Please look at fault P1469, same thing. Ford just needed to fill some space with more faults. **P1473** - Fan circuit open (VLCM) While the engine is running, the PCM detected a short/open circuit to the power to cooling fan.
P1474 - Fan control primary circuit While the engine is running, the PCM detected the primary fan control voltage was above or below its designated spec.
P1475 - Fan relay (low) circuit While the engine is running, the PCM detected the fan relay voltage was below its designated spec.
P1476 - Fan relay (high) circuit While the engine is running, the PCM detected the fan relay voltage was above its designated spec.

P1477 - Additional fan relay circuit While the engine is running, the PCM detected the additional fan relay voltage was above or below its designated spec.
P1478 - Cooling fan driver While the engine is running, the PCM detected the cooling fan driver voltage was above or below its designated spec.
P1479 - High fan control primary circuit While the engine is running, the PCM detected the high fan control voltage was above or below its designated spec.
P1480 - Fan secondary low with low fan on While the engine is running, the PCM detected the secondary fan to be below its designated spec.
P1481 - Fan secondary low with high fan on While the engine is running, the PCM detected the secondary fan to be above its designated spec.
P1483 - Fan circuit shorted to ground (VLCM) While the engine is running, the PCM detected a short circuit to the fan circuit
P1484 - Fan driver circuit open to power ground (VLCM) While the engine is running, the PCM detected an open circuit to the fan circuit
P1485 - EGRV circuit While the engine is running, the PCM detected that there is a fault in the EGRV
P1486 - EGRA circuit While the engine is running, the PCM detected that there is a fault in the EGRA
P1487 - Exhaust gas recirculation check solenoid circuit While the engine is running, the PCM detected a short/open circuit to the EGR check solenoid.
P1490 - Secondary air relief solenoid circuit While the engine is running, the PCM detected the secondary air relief solenoid voltage was above or below its designated spec.
P1491 - Secondary switch solenoid circuit While the engine is running, the PCM detected the secondary switch solenoid circuit voltage was above or below its designated spec.
P1500 - Vehicle speed sensor While the engine is running and the vehicle was moving, the PCM detected that the VSS signal was intermittent. Check the Real Time data, and look for intermittent changes in the output.

P1501 - Vehicle speed sensor out of self test range While the engine is running and the vehicle was moving, the PCM detected VSS the voltage was above or below its designated spec.
P1502 - Vehicle speed sensor intermittent While the engine is running and the vehicle was moving, the PCM detected the VSS to have an intermittent fault. Check the Real Time data, and look for intermittent changes in the output.
P1503 - Auxiliary speed sensor While the engine is running, the PCM detected the secondary VSS voltage was above or below its designated spec.
P1504 - Idle air control circuit While the engine is running for one minute, the PCM detected an electrical load failure on the IAC motor.
P1505 - Idle air control system at adaptive clip While the engine is running for one minute, the PCM detected the idle speed to be above its designated spec stored in the memory.
P1506 - Idle air control over speed error While the engine is running for one minute, the PCM detected the idle speed to be above its target idle speed.
P1507 - Idle air control under speed error While the engine is running for one minute, the PCM detected the idle speed to be below its target idle speed.
P1508 - Idle air control circuit open While the engine is running, the PCM detected an open circuit to the IAC
P1509 - Idle air control circuit shorted While the engine is running, the PCM detected a short circuit to the IAC
P1510 - Idle signal circuit While the engine is running, the PCM detected the IAC voltage was above or below its designated spec.
P1511 - Idle switch (electric control throttle) circuit While the engine is running, the PCM detected the Idle switch to be out of its designated spec
P1512 - Intake manifold runner control stuck closed (bank 1) While the engine is running, the PCM detected the IMRC bank 1 to be out of its designated spec. Check to see if the valve is stuck shut.

P1513 - Intake manifold runner control stuck closed (bank 2) While the engine is running, the PCM detected the IMRC bank 2 to be out of its designated spec. Check to see if the valve is stuck shut.
P1516 - Intake manifold runner control input error (bank 1) While the engine is running, the PCM detected the IMRC bank 1 monitor was above or below its designated spec.
P1517 - Intake manifold runner control input error (bank 2) While the engine is running, the PCM detected the IMRC bank 2 monitor was above or below its designated spec.
P1518 - Intake manifold runner control stuck open (bank 1) While the engine is running, the PCM detected the IMRC bank 1 to be out of its designated spec. Check to see if the valve is stuck open.
P1519 - Intake manifold runner control stuck closed (bank 2) While the engine is running, the PCM detected the IMRC bank 2 to be out of its designated spec. Check to see if the valve is stuck shut.
P1520 - Intake manifold runner control circuit While the engine is running, the PCM detected a short/open circuit to the IMRC.
P1521 - Variable resonance induction system solenoid #1 circuit While the engine is running, the PCM detected that there is a fault in the VRI solenoid 1.
P1522 - Variable resonance induction system solenoid #2 circuit While the engine is running, the PCM detected that there is a fault in the VRI solenoid 2.
P1523 - IVC solenoid circuit While the engine is running, the PCM detected that there is a fault in the IVC solenoid.
P1524 - Variable intake solenoid circuit While the engine is running, the PCM detected that there is a fault in the IVC solenoid.
P1525 - Air bypass valve While the engine is running, the PCM detected that there is a fault in the Air bypass valve.
P1526 - Air bypass system While the engine is running, the PCM detected that there is a fault in the Air bypass system.
P1527 - Accelerate warm-up solenoid circuit While the engine is running, the PCM detected that there is a fault in the accelerate warm-up solenoid.

P1528 - Subsidiary throttle valve solenoid circuit While the engine is running, the PCM detected that there is a fault in the SCAIR solenoid.
P1529 - SCAIR solenoid circuit While the engine is running, the PCM detected that there is a fault in the SCAIR circuit.
P1530 - A/C clutch circuit open While the engine is running, the PCM detected a short/open circuit in the power circuit to the clutch.
P1531 - Invalid test - accelerator pedal movement The accelerator pedal movement has failed its monitor test
P1533 - Air assisted injector circuit While the engine is running, the PCM detected the air assist injector voltage was above or below its designated spec.
P1534 - Restraint deployment indicator circuit While the engine is running, the PCM detected the SRS indicator to be out of its designated spec
P1537 - Intake manifold runner control stuck open (bank 1) While the engine is running, the PCM detected the IMRC bank 1 to be out of its designated spec. Check to see if the valve is stuck open.
P1538 - Intake manifold runner control stuck open (bank 2) While the engine is running, the PCM detected the IMRC bank 2 to be out of its designated spec. Check to see if the valve is stuck open.
P1539 - A/C clutch circuit over current/short While the engine is running, the PCM detected a short/open circuit to the A/C clutch circuit
P1540 - Air bypass valve circuit While the engine is running, the PCM detected that there is a fault in the air bypass valve.
P1549 - Intake manifold communication control circuit (bank 1) While the engine is running, the PCM detected that there is a fault in the IMCC.
P1550 - Power steering pressure sensor out of self test range While the engine is running, the PCM detected the Power steering pressure sensor voltage was above or below its designated spec.
P1562 - PCM B+ voltage low While the engine is running, the PCM detected the battery power is below its designated spec.
P1565 - Speed control command switch out of range high While the engine is running, the PCM detected that there is a fault in the VSS.

P1566 - Speed control command switch out of range low While the engine is running, the PCM detected that there is a fault in the VSS.
P1567 - Speed control output circuit While the engine is running, the PCM detected that there is a fault in the VSS.
P1568 - Speed control unable to hold speed While the engine is running, the PCM detected that there is a fault in the VSS.
P1569 - Intake manifold runner control output circuit low While the engine is running, the PCM detected the IMRC to be below its designated spec.
P1570 - Intake manifold runner control output circuit high While the engine is running, the PCM detected the IMRC to be above its designated spec.
P1571 - Brake switch While the engine is running, the PCM detected that there is a fault in the brake pedal switch.
P1572 - Brake pedal switch circuit While the engine is running, the PCM detected the brake pedal switch to be out of its designated spec.
P1573 - Throttle position not available While the engine is running, the PCM detected TPS is not available. Check the Real Time data, and look for intermittent changes in the output.
P1574 - Throttle position sensor outputs disagree While the engine is running, the PCM detected TPS outputs disagree. Check the Real Time data, and look for intermittent changes in the output.
P1575 - Pedal position out of self test range While the engine is running, the PCM detected that there is a fault in the PPS.
P1576 - Pedal position not available While the engine is running, the PCM detected PPS is not available. Check the Real Time data, and look for intermittent changes in the output.
P1577 - Pedal position sensor outputs disagree While the engine is running, the PCM detected PPS outputs disagree. Check the Real Time data, and look for intermittent changes in the output.
P1578 - ETC power less than demand While the engine is running, the PCM detected the ETC to be below its designated spec.
P1579 - ETC in power limiting mode While the engine is running, the PCM detected that there is a fault in the ETC.

P1580 - Electronic throttle monitor PCM override
While the engine is running, the PCM detected that there is a fault in the electronic throttle. The PCM has put the engine in fail-safe mode.

P1581 - Electronic throttle monitor malfunction
While the engine is running, the PCM detected that there is a fault in the electronic throttle.

P1582 - Electronic throttle monitor data available
While the engine is running, the PCM detected that there is a fault in the electronic throttle.

P1583 - Electronic throttle monitor cruise disablement
While the engine is running, the PCM detected that there is a fault in the electronic throttle. The PCM has disabled the cruise control.

P1585 - Throttle control malfunction
While the engine is running, the PCM detected that there is a fault in the throttle control.

P1586 - Electronic throttle to PCM communication error
While the engine is running, the PCM detected that there is a fault in the electronic throttle. Check the CAN form the ET to the PCM.

P1600 - Loss of KAM power, circuit open
While the engine is running, the PCM detected an open circuit. Check fuses, or any power source to the PCM.

P1601 - ECM/TCM serial communication error
While the engine is running, the PCM detected the CAN communication between the ECM and TCM was above or below its designated spec.

P1602 - Immobilizer/ECM communication error
While the engine is running, the PCM detected the CAN communication between the ECM and Immobilizer was above or below its designated spec.

P1605 - Keep alive memory test failure
While the engine is running, the PCM detected an open circuit. Check fuses, or any power source to the PCM. When the PCM looses its KAM, this fault is set.

P1606 - ECM control relay output circuit
While the engine is running, the PCM detected that there is a fault in the ECM control relay. Check relay for operation.

P1607 - MIL output circuit While the engine is running, the PCM detected that there is a fault in the MIL circuit. Check if the MIL bulb is blown, also check the power and ground.
P1608** - PCM internal circuit
P1609** - Diagnostic lamp driver
P1610** - SBDS interactive codes
P1611** - SBDS interactive codes
P1612** - SBDS interactive codes
P1613** - SBDS interactive codes
P1614** - SBDS interactive codes
P1615** - SBDS interactive codes
P1616** - SBDS interactive codes
P1617** - SBDS interactive codes
P1618** - SBDS interactive codes
P1619** - SBDS interactive codes
P1620** - SBDS interactive codes
P1621** - Immobilizer code words do not match
P1622** - Immobilizer ID does not match
P1623** - Immobilizer code word/ID number write failure
P1624 - Anti-theft system While the engine is running, the PCM detected that there is a fault in the anti-theft system
P1625 - Fan driver circuit open to power B+ (VLCM) While the engine is running, the PCM detected that there is a fault in the VLCM. Check power supply to the VLCM fan circuit.
P1626 - A/C circuit open to power B+ (VLCM) While the engine is running, the PCM detected that there is a fault in the VLCM. Check power supply to the VLCM fan circuit.
P1627 - Module supply voltage out of range While the engine is running, the PCM detected the module supply voltage was above or below its designated spec.

****Note:** Form P1608 to P1623 faults are internal PCM faults. Possible PCM failure, or improper PCM coding.

P1628 - Module ignition supply input While the engine is running, the PCM detected the module ignition voltage was above or below its designated spec.
P1629 - Internal voltage regulator While the engine is running, the PCM detected that there is a fault in the alternator.
P1630 - Alternator regulator #1 control circuit While the engine is running, the PCM detected that there is a fault in the alternator.
P1631 - Alternator regulator #2 control circuit While the engine is running, the PCM detected that there is a fault in the alternator.
P1632 - Smart alternator faults sensor/circuit While the engine is running, the PCM detected that there is a fault in the alternator.
P1633 - Keep alive power voltage too low While the engine is running, the PCM detected an open circuit. Check fuses, or any power source to the PCM. When the PCM looses its KAM, this fault is set.
P1634 - Data output link circuit While the engine is running, the PCM detected that there is a fault in the CAN.
P1635 - Tire/axle out of acceptable range While the engine is running, the PCM detected an internal fault. Check if the tires and rims are all the same size, and if they are meant to be on the vehicle.
P1636 - Inductive signature chip communication error While the engine is running, the PCM detected an internal fault.
P1637 - CAN link ECM/ABSCM circuit/network While the engine is running, the PCM detected a fault in the CAN communication between the ECM and the ABS control unit.
P1638 - CAN link ECM/INSTM circuit/network While the engine is running, the PCM detected a fault in the CAN communication between the ECM and the Instrument cluster
P1639 - Vehicle ID block corrupted, not programmed While the engine is running, the PCM detected an internal fault.
P1640 - Powertrain DTCs available in another control module (ref. PID 0946) This one is an interesting fault. If you see this fault, check to see if you have any other control modules responding. If so, please check faults in that control module.

P1641 - Fuel pump primary circuit While the engine is running, the PCM detected the FPDM circuit to be out of its designated spec
P1642 - Fuel pump monitor circuit high input While the engine is running, the PCM detected the FPDM to be above its designated spec.
P1643 - Fuel pump monitor circuit low input While the engine is running, the PCM detected the FPDM to be below its designated spec.
P1644 - Fuel pump speed control circuit While the engine is running, the PCM detected that there is a fault in the FPDM
P1645 - Fuel pump resistor switch circuit While the engine is running, the PCM detected that there is a fault in the FPDM
P1646 - Linear O2 sensor control chip (bank 1) While the engine is running, the PCM detected an internal fault.
P1647 - Linear O2 sensor control chip (bank 2) While the engine is running, the PCM detected an internal fault.
P1648 - Knock sensor input chip While the engine is running, the PCM detected an internal fault.
P1649 - Fuel injection pump module While the engine is running, the PCM detected an internal fault.
P1650 - Power steering pressure switch out of self test range While the engine is running, the PCM detected the PSPS voltage was above or below its designated spec.
P1651 - Power steering pressure switch input While the engine is running, the PCM detected that there is a fault in the PSPS
P1652 - Idle air control monitor disabled by PSPS failed on While the engine is running, the PCM detected that there is a fault in the PSPS
P1653 - Power steering output circuit While the engine is running, the PCM detected that there is a fault in the PSPS
P1654 - Recirculation override circuit While the engine is running, the PCM detected that there is a fault in the recirculation override circuit
P1655 - Starter disable circuit While the engine is running, the PCM detected that there is a fault in the starter disable circuit. Check the neutral safety switch.

P1656 - CAN link PCM/PCM circuit/network While the engine is running, the PCM detected that there is a fault in the CAN network.
P1657 - CAN link chip malfunction While the engine is running, the PCM detected an internal fault.
P1660 - Output circuit check circuit high input While the engine is running, the PCM detected an internal fault.
P1661 - Output circuit check circuit low input While the engine is running, the PCM detected an internal fault.
P1662 - EDU_EN Output circuit While the engine is running, the PCM detected an internal fault.
P1663 - Fuel demand command signal output circuit While the engine is running, the PCM detected an internal fault.
P1667 - Cylinder ID circuit While the engine is running, the PCM detected an internal fault.
P1668 - PCM/IDM communications error While the engine is running, the PCM detected an internal fault.
P1670 - Electronic feedback signal not detected While the engine is running, the PCM detected an internal fault.
P1680 - Metering oil pump failure While the engine is running, the PCM detected that there is a fault in the MOP
P1681 - Metering oil pump failure While the engine is running, the PCM detected that there is a fault in the MOP
P1682 - Metering oil pump failure While the engine is running, the PCM detected that there is a fault in the MOP
P1683 - Metering oil pump temperature sensor circuit While the engine is running, the PCM detected the MOP temp sensor voltage was above or below its designated spec.
P1684 - Metering oil pump position sensor circuit While the engine is running, the PCM detected that there is a fault in the MOP.
P1685 - Metering oil pump stepping motor cont. circuit While the engine is running, the PCM detected that there is a fault in the MOP.

P1686 - Metering oil pump stepping motor cont. circuit While the engine is running, the PCM detected that there is a fault in the MOP.
P1687 - Metering oil pump stepping motor cont. circuit While the engine is running, the PCM detected that there is a fault in the MOP.
P1688 - Metering oil pump stepping motor cont. circuit While the engine is running, the PCM detected that there is a fault in the MOP.
P1689 - Oil pressure control solenoid circuit While the engine is running, the PCM detected that there is a fault in the MOP.
P1690 - Wastegate solenoid circuit While the engine is running, the PCM detected the Wastegate solenoid voltage was above or below its designated spec.
P1691 - Turbo pressure control solenoid circuit While the engine is running, the PCM detected the turbo pressure control solenoid voltage was above or below its designated spec.
P1692 - Turbo control solenoid circuit While the engine is running, the PCM detected the turbo control solenoid voltage was above or below its designated spec.
P1693 - Turbo charge control circuit While the engine is running, the PCM detected the turbo charge circuit voltage was above or below its designated spec.
P1694 - Turbo charge relief circuit While the engine is running, the PCM detected the turbo charge relief circuit voltage was above or below its designated spec.
P1700*** - Transmission indeterminate failure (failed to neutral)
P1701*** - Reverse engagement error
P1702*** - Transmission range sensor circuit intermittent
P1703*** - Brake switch out of self-test range
P1704*** - Transmission range circuit not indicating park/neutral during self-test
P1705*** - Transmission range circuit not indicating park/neutral during self-test
P1705*** - Throttle position sensor (A/T)
P1706*** - High vehicle speed observed in park

*****Note:** Form fault codes P1700 to P1877 all refer to Transmission or transfer case problems.
With the wide variety of transmissions and transfer cases, please refer to the manual of that unit.

P1707*** - Transfer case neutral indicator hard fault present
P1708*** - Clutch switch circuit
P1709*** - Park neutral position switch out of self-test range
P1710*** - TCM solenoid/internal ground circuit
P1711*** - Transmission fluid temperature sensor out of self-test range
P1712*** - Transmission torque reduction request signal
P1713*** - Transmission fluid temperature sensor in range failure (<50 °F)
P1714*** - Shift solenoid A inductive signature
P1715*** - Shift solenoid B inductive signature
P1716*** - Shift solenoid C inductive signature
P1717*** - Shift solenoid D inductive signature
P1718*** - Transmission fluid temperature sensor in range failure (>250 °F)
P1719*** - Engine torque signal
P1720*** - Vehicle speed (meter) circuit
P1721*** - Gear 1 incorrect ratio
P1722*** - Gear 2 incorrect ratio
P1722*** - Stall speed
P1723*** - Gear 3 incorrect ratio
P1724*** - Gear 4 incorrect ratio
P1725*** - Insufficient engine speed increase during self test
P1726*** - Insufficient engine speed decrease during self test
P1726*** - Engine over speed
P1727*** - Coast clutch solenoid inductive signature
P1728*** - Transmission slip
P1729*** - 4x4L switch
P1730*** - Gear control malfunction 2,3,5
P1731*** - 1-2 shift malfunction
P1731*** - Inconsistent gear ratio

*****Note:** Form fault codes P1700 to P1877 all refer to Transmission or transfer case problems. With the wide verity of transmissions and transfer cases, please refer to the manual of that unit.

P1732*** - 2-3 shift malfunction
P1733*** - 3-4 shift malfunction
P1734*** - 4-5 shift malfunction
P1735*** - First gear switch circuit failure
P1736*** - Second gear switch circuit failure
P1737*** - Lockup solenoid
P1738*** - Shift time error
P1739*** - Slip solenoid
P1740*** - Torque converter clutch solenoid inductive signature
P1741*** - Torque converter clutch solenoid control error
P1742*** - Torque converter clutch solenoid circuit failed on
P1743*** - Torque converter clutch solenoid circuit
P1744*** - Torque converter clutch solenoid circuit
P1745*** - Line pressure solenoid
P1746*** - Pressure control solenoid A open circuit
P1747*** - Pressure control solenoid A short circuit
P1747*** - Pressure regulator 3
P1748*** - Pressure control solenoid
P1748*** - Pressure regulator 5
P1749*** - Pressure control solenoid A failed low
P1751*** - Shift solenoid A performance
P1752*** - Shift solenoid A circuit short
P1754*** - Coast clutch solenoid circuit
P1756*** - Shift solenoid B performance
P1756*** - Shift solenoid B circuit open
P1757*** - Shift solenoid B circuit short
P1760*** - Pressure control solenoid A short circuit intermittent
P1761*** - Shift solenoid C performance

*****Note:** Form fault codes P1700 to P1877 all refer to Transmission or transfer case problems. With the wide variety of transmissions and transfer cases, please refer to the manual of that unit.

P1762*** - Overdrive band failed off
P1765*** - Timing solenoid circuit
P1766*** - Shift solenoid D performance
P1767*** - Torque converter clutch circuit
P1768*** - Performance/normal/winter mode input
P1769*** - AG4 transmission torque modulation fault (VW trans)
P1770*** - Clutch solenoid circuit
P1771*** - Shift solenoid E performance
P1772*** - Throttle position sensor circuit low input
P1775*** - Transmission system MIL fault
P1776*** - Ignition retard request duration
P1777*** - Ignition retard request circuit
P1778*** - Transmission reverse I/P circuit
P1779*** - TCIL circuit
P1780*** - Transmission control switch (O/D cancel) circuit out of self test range
P1781*** - 4x4L circuit out of self test range
P1782*** - Performance/economy switch circuit out of self test range
P1783*** - Transmission over temperature condition
P1784*** - Transmission mechanical failure - first and reverse
P1785*** - Transmission mechanical failure - first and second
P1786*** - 3-2 downshift error
P1787*** - 2-1 downshift error
P1788*** - Pressure control solenoid B open circuit
P1789*** - Ignition supply malfunction >7, >9 volts
P1790*** - TP (mechanical) circuit
P1791*** - TP (electric) circuit
P1792*** - Barometer pressure circuit
P1793*** - Intake air volume circuit

***Note: Form fault codes P1700 to P1877 all refer to Transmission or transfer case problems. With the wide variety of transmissions and transfer cases, please refer to the manual of that unit.

P1794*** - Battery voltage circuit
P1795*** - Idle switch circuit
P1796*** - Kick down switch circuit
P1797*** - Clutch pedal position switch/neutral switch circuit
P1798*** - Coolant temperature circuit
P1799*** - Hold switch circuit
P1804*** - 4-wheel drive high indicator circuit open or shorted to ground
P1806*** - 4-wheel drive high indicator short to battery
P1808*** - 4-wheel drive low indicator circuit open or short to ground
P1810*** - 4-wheel drive low indicator short to battery
P1812*** - 4-wheel drive mode select switch circuit open
P1815*** - 4-wheel drive mode select switch circuit short to ground
P1819*** - Neutral safety switch input short to ground
P1820*** - Transfer case LO to HI shift relay circuit open or short to ground
P1822*** - Transfer case LO to HI shift relay coil short to battery
P1824*** - 4-wheel drive electric clutch relay open or short to ground
P1826*** - 4-wheel drive electric clutch relay short to battery
P1828*** - Transfer case HI to LO shift relay coil circuit open or short to ground
P1830*** - Transfer case HI to LO shift relay coil circuit short to battery
P1832*** - Transfer case 4-wheel drive solenoid circuit open or short to ground
P1834*** - Transfer case 4-wheel drive solenoid circuit short to battery
P1838*** - No shift motor movement detected
P1846*** - Transfer case contact plate 'A' circuit open
P1850*** - Transfer case contact plate 'B' circuit open
P1854*** - Transfer case contact plate 'C' circuit open
P1858*** - Transfer case contact plate 'D' circuit open
P1866*** - Transfer case cannot be shifted
P1867*** - Transfer case contact plate general circuit failure

*****Note:** Form fault codes P1700 to P1877 all refer to Transmission or transfer case problems.
With the wide variety of transmissions and transfer cases, please refer to the manual of that unit.

P1876*** - Transfer case 2-wheel drive solenoid circuit open or short to ground
P1877*** - Transfer case 2-wheel drive solenoid circuit short to battery
P1881 - Engine coolant level switch circuit While the engine is running, the PCM detected the ECL switch to be out of its designated spec.
P1882 - Engine coolant level switch circuit short to ground While the engine is running, the PCM detected a short circuit to the ECL switch.
P1883 - Engine coolant level switch circuit While the engine is running, the PCM detected the ECL switch to be out of its designated spec.
P1884 - Engine coolant level lamp circuit short to ground While the engine is running, the PCM detected a short circuit to the ECL switch.
P1891 - Transfer case contact plate ground return open circuit While the engine is running, the PCM detected a short to ground circuit to the transfer case contact plate.
P1900 - Output shaft speed sensor circuit intermittent While the engine is running, the PCM detected the output shaft speed sensor to be intermittent. Check the Real Time data, and look for intermittent changes in the output.
P1901 - Turbine shaft speed sensor circuit intermittent While the engine is running, the PCM detected torque converter speed sensor to be intermittent. Check the Real Time data, and look for intermittent changes in the output.
P1906 - Kickdown pull relay open or short circuit to ground (A4LD) While the engine is running, the PCM detected a short/open circuit to the kickdown relay.
P1907 - Kickdown hold relay open or short circuit to ground (A4LD) While the engine is running, the PCM detected a short/open circuit to the kickdown relay.
P1908 - Transmission pressure control solenoid open or short (A4LD) While the engine is running, the PCM detected a short/open circuit to the trans pressure control solenoid.
P1909 - Transmission fluid temperature sensor circuit open or shorted (A4LD) While the engine is running, the PCM detected a short/open circuit to the trans fluid temp sensor.

*****Note:** Form fault codes P1700 to P1877 all refer to Transmission or transfer case problems.
With the wide variety of transmissions and transfer cases, please refer to the manual of that unit.

10

With this chapter, you will take your scan tool to the next level for Daimler-Chrysler vehicles. Even when you know how to use the tool to interrogate faults, Daimler-Chrysler, like every other manufacturer, has its own ins and outs.

Chapter 10 > OBDII Diagnostic Tips for Daimler-Chrysler

ENTER

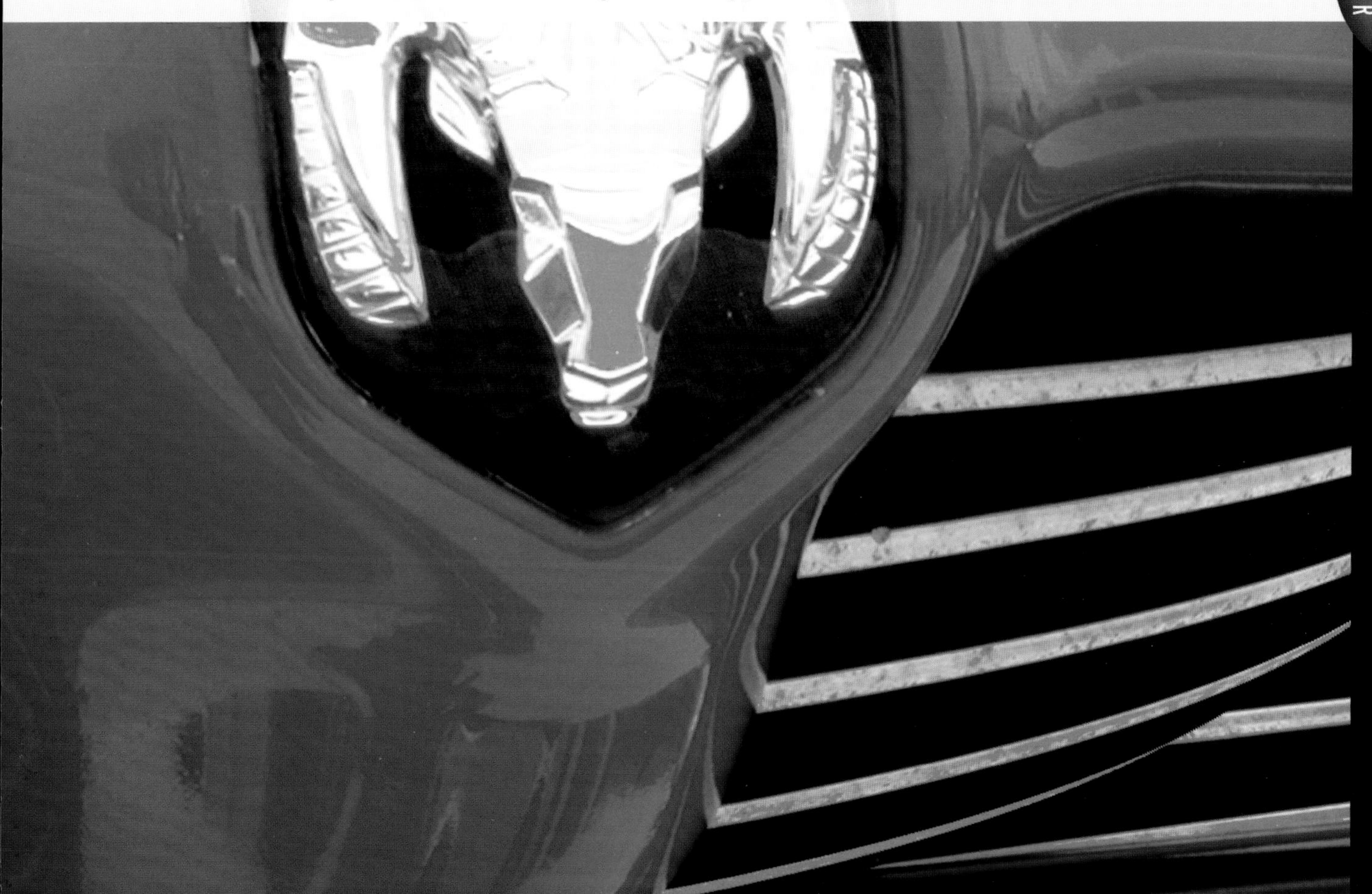

General Information about Chrysler

We all know that Mercedes Benz bought Chrysler/Jeep. So please forgive us if we refer to Daimler-Chrysler as Chrysler from now on.

In 1995, Chrysler began to phase in the OBDII system on the Neon and Eagle Talon models that were equipped with a 2.0L engine. The OBDII systems that were used on these vehicles included the Misfire Monitor. Chrysler had this operate with a lower threshold Misfire Detection System designed to monitor misfires without setting any codes. Also, the EVAP Monitor did not have to operate on these vehicles. As you can see, a lot was missing with these systems.

In 1996 and beyond, Chrysler had to comply with the OBDII requirements. These requirements had to apply to both gas and diesel vehicles.

Understanding Chrysler Diagnostics

Chrysler's control units

Chrysler uses a Single Board Engine Controller (or SBEC). This system, controls the Engine and Transmission. In addition, it meets all OBDII requirements.

Diagnostic Task Manager

Chrysler uses the Diagnostic Task Manager to detect if components are operating properly. This software is in the engine control module. It organizes and prioritizes the emission control tests (or Monitor tests) and it records and displays the test results of the diagnostic procedures. These monitors are run under specific operation conditions, called enable criteria. For the enable criteria to happen, a Drive Cycle must take place.

Chrysler's Drive Cycle

The purpose of the Drive Cycle is to run all the OBDII monitor tests. While on the road test for the Drive Cycle, connect a scan tool the to DLC. Have an assistant watch the Readiness status of the Non-Continuous and Continuous Monitors.

Chrysler Specific Monitor Tests

Catalyst Monitor

Purpose To monitor the catalysts ability to accumulate and discharge oxygen needed to complete the emissions reducing chemical reactions. As the CAT ages, its ability to store oxygen is lessened. This monitor is to check the CATs ability to store and discharge oxygen.

Test Procedure Warm the engine to normal operating temperature (Check the ECT Real Time Data, if over - 147 °F). Accelerate to 40 to 55 MPH, with engine RPM at 1200 to 1700. After three minutes, check the Real Time Data value, the Intake manifold absolute pressure should be at 10.5 to 15.0 in.hg. If not, check the condition of the CAT or Oxygen Sensors. In addition, a fault can be stored, if a new Rear Oxygen sensor is installed with an aging Front Oxygen sensor.

EGR System Monitor

Purpose To test the reliability and flow characteristics of the EGR system.

Test Procedure Warm the engine to normal operating temperature (Check the ECT Real Time Data, if over - 180 °F). Rev engine speed to 2248-2688 RPM (for automatic transmission) or 1952-2400 RPM (for manual transmission). Check the Real Time Data value for Short Term Fuel trim. It should be less than +4.4% or greater than -8%, and Vehicle speed input should be more than 10 MPH. Then, have an assistant check a scan tool while accelerating through all the gears and achieve a steady speed from 25 to 50 MPH. After one minute, the EGR system monitor should complete. The maximum for this test to complete is four minutes.

Note: The EGR System Monitor will be suspended if the Fuel Trim, Oxygen Sensor, Oxygen Sensor Heater or Misfire Monitor is running. In addition, the EGR system monitor can have conflicts. The Task Manager will not run the EGR System Monitor if any of these conditions is present:

- CAT, EVAP or Fuel (rich intrusive test) Monitor in progress
- Fuel System Monitor 1-Trip pending rich or lean code is stored
- Oxygen Sensor or Heated Monitor 1-Trip pending code is stored
- Misfire Monitor 1-Trip pending code is stored

EVAP monitor

Purpose To test the fuel tank and fuel delivery systems. The EVAP system prevents fuel tank vapors from entering the atmosphere. Fuel evaporation releases Hydrocarbons (HC). HC destroys the ozone.
EVAP Types:

- Purge Flow Monitor
This monitor is used on less strict evaporative systems to identify the ability of purge flow throughout the system by a measurement of change in IAC stepper motor position, engine speed or the average O2 controller. This monitor can be tested under various engine operating conditions including idle.
- Leak Detection Pump (LDP) System Monitor
A vacuum motor with a control valve runs to create a test pressure in the evap system of approximately 7" of water. The engine control unit measures pump run time to see if there is a leak. If it runs for too short a period then the engine control unit assumes a restriction is present. If the pump runs for too long a period, then the engine control unit assumes that there is a leak.
- Leak Detection Pump (LDP) System w/ Proportional Purge Solenoid
This system is the same as the LDP system, but it incorporates the Proportional Purge Solenoid. This solenoid is controlled by the PCM and operates at 200 Hz. The engine control unit monitors the current being applied to the solenoid and then adjusts that current to achieve the desired purge flow. The solenoid then controls the purge flow of the fuel vapors from the vapor canister and fuel tank to the intake manifold. The engine control unit shuts off the solenoid

when the cold start warm-up and the hot start time delay happen.

Test Procedure Before the test, the engine must be between 40 to 90 °F and the fuel level must be at least 15% or over. Then, have an assistant check a scan tool while accelerating through all the gears and achieve a steady speed from 25 to 50 MPH. After one minute, the EVAP system monitor should complete. The maximum for this test to complete is four minutes. Things to remember: This monitor will set a pending code if it detects low airflow through the system. If this happens for two consecutive trips, the MIL and the code will be set to active. Any restrictions or disconnected lines within the EVAP system vacuum lines or a faulty solenoid can set a code too.

All Other Monitors

The fuel level must be from 15-85% to start the test.

The ambient temperature must be from 40-90 °F (if the temperature drops below 50 °F during the Drive Cycle, the EVAP monitor may not complete its test).
Allow the engine to fully warm up to 170 °F prior to the test. After all the above have been accomplished:

- Turn the Ignition off for 10 seconds
- Restart engine, and let idle for two minutes
- Accelerate to 30 MPH for 2 minutes, then to over 40 MPH and hold the throttle steady for over 3 minutes
- Decelerate to idle speed and then turn the Ignition key off for 10 seconds.

If you can't find a road that will let you accomplish the above road test then:

- Drive at a speed of over 45 MPH for 4 to 5 minutes, and then turn the ignition key off.

Daimler-Chrysler specific Fault codes (P1)

P1110 - Decrease Engine Performance Due to High Intake Air Temperature.
This fault is mainly used for engines with a turbo, or supercharger. If the engine has a turbo, look for leaks from the exhaust to the intake on the turbo, also, the intercooler might be malfunctioning.

P1180 - Decreased Engine Performance Due to High Injection Pump Fuel Temp.
This fault will is only used with diesel engines. Most injection pumps are oil cooled and lubricated, so check the oil flow.

P1193 - IAT input high
Check for fault in wiring, and in the sensor itself, refer to the real time data to check sensor input

P1195 - 1/1 02 Sensor Slow During Catalyst Monitor
This is usually due to a lazy oxygen sensor. Check the sensor with real time data, and the circuit for faulty wiring

P1196 - 2/1 02 Sensor Slow During Catalyst Monitor
This is usually due to a lazy oxygen sensor. Check the sensor with real time data, and the circuit for faulty wiring

P1197 - 1/2 02 Sensor Slow During Catalyst Monitor
This is usually due to a lazy oxygen sensor. Check the sensor with real time data, and the circuit for faulty wiring

P1198 - Radiator Temperature Sensor Volts Too High.
Check coolant system for flow problem

P1199 - Radiator Temperature Sensor Volts Too Low
Check coolant system for flow problem

P1281 - Engine is cold Too Long
Check the thermostat, not opening properly, possible stuck, or radiator is clogged.

P1282 - Fuel Pump Relay Control Circuit.
Check voltage to fuel pump rely. Must be over 10 volts. If not check fuse, or PCM. If over 10 volts, check pump

P1283 - Idle Select Signal Invalid.

P1284 - Fuel Injection Pump Battery Voltage Out-of-Range.

P1285 - Fuel Injection Pump Controller Always On, Fuel injection pump module relay circuit failure detected.

P1286 - Accelerator Position Sensor (APPS) Supply Voltage Too High.

P1287 - Fuel Injection Pump Controller Supply Voltage Low.

P1288 - Intake Manifold Short Runner Solenoid Circuit, An open or shorted condition detected in the short runner tuning valve circuit.

P1289 - Manifold Tune Valve Solenoid Circuit, An open or shorted condition detected in the manifold tuning valve solenoid control circuit.

P1290 - CNG Fuel System Pressure Too High.

P1291 - No Temp Rise Seen From Intake Heaters

P1292 - CNG Pressure Sensor Voltage Too High

P1293 - CNG Pressure Sensor Voltage Too Low.

P1294 - Target Idle Not Reached. Possible vacuum leak or IAC (AIS) lost steps.

P1295 - Accelerator Position Sensor (APPS) Supply Voltage Too Low.

P1296 - Loss of a 5 volt feed to the MAP Sensor has been detected.

P1297 - No Change in MAP From Start to Run.

P1298 - Lean Operation at Wide Open Throttle.

P1299 - Vacuum Leak Found (IAC Fully Seated).Possible vacuum leak.

P1388 - An open or shorted condition detected in the ASD or CNG shutoff relay control circuit.

P1388 - An open or shorted condition detected in the auto shutdown relay circuit.

P1389 - No ASD Relay Output Voltage at PCM.

P1390 - Timing Belt Skipped 1 Tooth or More

P1391 - Intermittent Loss of CMP or CKP.

P1398 - Miss-Fire Adaptive Numerator at Limit.

P1399 - Wait To Start Lamp circuit.

P1403 - No 5V to EGR Sensor.

P1476 - Aux 5 Volt Supply Voltage High.

P1477 - Too Much Secondary Air.
P1478 - Battery Temp Sensor Volts Out of Limit.
P1479 - Transmission Fan Relay Circuit.
P1480 - An open or shorted condition detected in the PCV solenoid circuit.
P1481 - EATX RPM pulse generator signal for misfire detection does not correlate with expected value.
P1482 - Catalyst Temperature Sensor Circuit Shorted Low.
P1483 - Catalyst Temperature Sensor Circuit Shorted High.
P1484 - Catalytic Converter Overheat Detected.
P1485 - Air Injection Solenoid Circuit.
P1486 - Evap Leak Monitor Pinched Hose Found.
P1487 - Hi Speed Radiator Fan CTRL Relay Circuit.
P1488 - Auxiliary 5 Volts Supply Output Too Low.
P1489 - High Speed Fan CTRL Relay Circuit.
P1490 - Low Speed Fan CTRL Relay Circuit.
P1491 - Radiator Fan control Relay Circuit.
P1492 - Ambient/Batt Temp Sensor Volts Too High.
P1493 - Ambient/Batt Temp Sensor Volts Too Low.
P1493 - Ambient/Batt Temp Sensor Volts Too Low.
P1494 - Leak Detection Pump Switch or Mechanical Fault.
P1495 - Leak Detection Pump solenoid Circuit.
P1496 - 5 Volt Supply, Output Too Low.
P1498 - High Speed Radiator Fan Ground CTRL Relay Circuit.
P1499 - Hydraulic cooling fan solenoid circuit.
P1594 - Charging System Voltage Too High.
P1595 - Speed Control Solenoid Circuits.
P1596 - Speed Control Switch Always High.

P1597 - Speed Control Switch Always Low.
P1598 - A/C Pressure Sensor Volts Too High.
P1599 - A/C Sensor Input Low.
P1680 - Clutch Released Switch Circuit.
P1681 - No I/P Cluster CCD/J1850 Messages Received.
P1682 - Charging System Voltage Too Low.
P1683 - Speed ctrl power relay, or s/c 12V driver circuit.
P1684 - Batt Lost in 50 Star.
P1685 - SKIM Invalid Key.
P1686 - No SKIM BUS Messages Received.
P1687 - No MIC BUS Message.
P1688 - Internal Fuel Injection Pump Controller Failure.
P1689 - No Communication between ECM and Injection Pump Module.
P1690 - Fuel Injection Pump CKP Sensor Does Not Agree With ECM CKP Sensor.
P1691 - Fuel Injection Pump Controller Calibration Error.
P1692 - A Companion DTC was set in both the ECM and PCM.
P1693 - DTC Detected in Companion Module.
P1694 - Fault In Companion Module.
P1695 - No CCD/J1850 Message From Body Control Module.
P1696 - PCM Failure EEPROM Write Denied.
P1697 - PCM Failure SRI Mile Not Stored.
P1698 - No CCD Message received from PCM.
P1719 - Skip Shift Solenoid Circuit.
P1740 - TCC OR O/D Solenoid Performance.
P1756 - Governor Pressure Not Equal to Target @ 15-20 PSI.
P1757 - Governor Pressure Above 3 PSI In Gear With O MPH

P1762 - Governor Press Sensor Offset Volts Too Low or High.
P1763 - Governor Pressure Sensor Volts Too Hi.
P1764 - Governor Pressure Sensor Volts Too Low.
P1765 - Trans 12 Volt Supply Relay Ctrl Circuit.
P1899 - P/N Switch Stuck in Park or in Gear.

11

EOBD means European On-Board Diagnostics. Europe, not to be outdone by the Americans, came up with their own Diagnostic standards. They did this out of necessity, since they had to make the vehicles manufactured after January 1996 OBDII compliant in order to be sold in the USA. They thought "Why make this exception only for the Americans? Let's only make one type of vehicle." And after 6 years, they did.

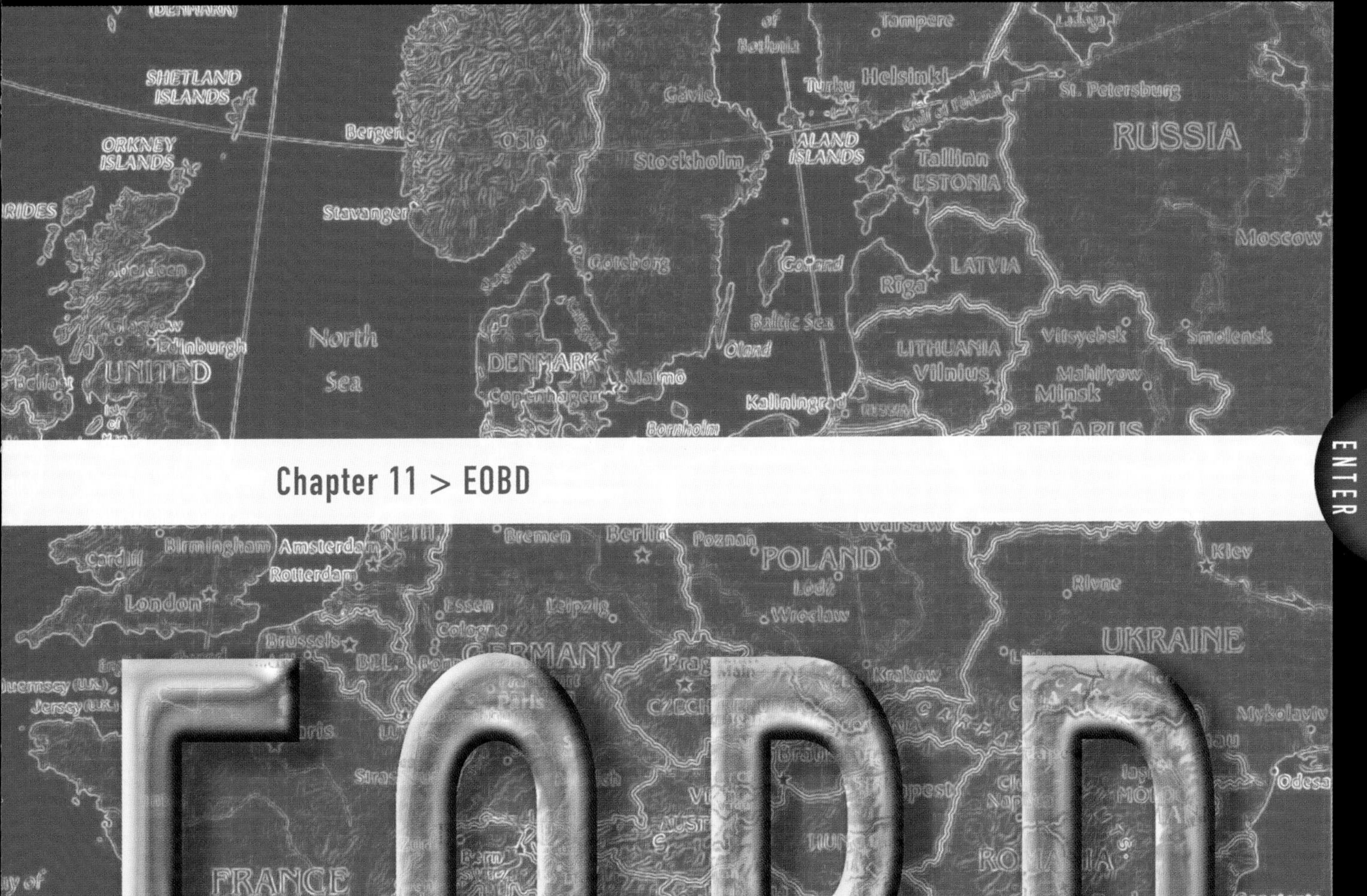

Chapter 11 > EOBD

ENTER

European On-Board Diagnostics or EOBD

OBDII is not a world wide standard. (Frankly, some Europeans think we are crazy with all the rules we have on our vehicles). However, since January of 2002, the Europeans are now following the standards that we proposed. They realized that OBDII is helping the environment, and the ozone. Moreover, they realized that it is too expensive to develop different vehicles for different markets. They developed a system based on the standards of the U.S. version of OBDII, called EOBD. Since they never had an OBDI, it makes sense that they wouldn't put a numeral after the OBD.

All gasoline vehicles manufactured and sold in Europe after January 1, 2002 are required to comply to EOBD. Soon, they will force all diesel vehicles to comply as well (that is many vehicles - in Europe every other car is diesel).

In the US, we have the SAE, which made most of the standards for OBDII. In Europe, they have an organization called the International Standard Organization (or ISO), which makes most of the standards. The ISO 9141 standard is for most Asian, European, some GM and some Chrysler vehicles. ISO 9141 is a part of OBDII specifications. Most Americans are already familiar with it. When ARB and EPA designed OBDII, they took standards that were in practice, and expanded upon them. So, don't think that OBDII is only an American thing.

EOBD is very similar to OBDII. EOBD drive cycle is a good example.

EOBD Drive Cycle

A drive cycle will be started by the ECM as soon as the engine is started (from hot or cold). The drive cycle ends when the engine is switched off. The drive cycle should include a warm up cycle. A complete drive cycle should allow the system to perform diagnostics on all areas and to run all monitors. For the technician or the EOBD system to perform a full drive cycle, all of the following conditions will need to be met:

- Cold start. For the ECM to recognize a cold start the coolant temperature must be below 50°C and within 6°C of the ambient air temperature at start up.

• Idle. The vehicle should be allowed to idle for around 2 - 3 minutes with an electrical load, such as the rear screen demister switched on. This will allow the oxygen sensor heater, the misfire monitor and, if closed loop fueling is achieved, the fuel trim monitor to run.

• The vehicle now needs to be accelerated up to a speed of around 50 - 60 miles per hour. During this acceleration, the misfire and fuel trim diagnostics will be performed.

• The vehicle speed should now be held steady at the above speed for around 3 minutes. The ECM will run the O2 response misfire and fuel trim monitors.

• At the end of this three-minute period, the accelerator pedal should be released and the vehicle allowed to slow to 20 miles per hour without changing gear or using the brakes or clutch. Fuel trim diagnostics will again be run.

• The vehicle should now be accelerated back up to the same speed as before - the same diagnostics as in step three above are run.

• The speed should be held again, this time for around five minutes. As well as performing the same diagnostics as in step three, the catalyst monitor is now run. If the catalyst efficiency is low or the battery has been disconnected it may take up to five complete drive cycles for the ECM to determine the effectiveness of the catalyst.

• Finally, the vehicle should be decelerated, as per step five, when the same tests will again be run.

Yep, the same strange requests as seen already with OBDII are also used with EOBD (I am still trying to find a stretch of road to perform these crazy road tests). An engine can only be monitored a certain way. Both systems have to have more or less the same monitors.

The major differences between OBDII and EOBD are the fault codes. The texts on the majority of the faults are different. No, it is not because of the languages, but people use different terms in different parts of the world. ISO has own set of terms.

In layman's terms, EOBD is the diagnostic standard of the European Community. OBDII and EOBD are very similar. EOBD is not as wide spread as OBDII yet, but it is gaining popularity.

12

What is in store for us in the future? Will there be an OBDIII? Questions like this, and more, will be answered in this chapter.

So, stay tuned to this chapter and find out. . .

Chapter 12 > Future of OBDII

What's Next?

Twenty-Five years ago, I first started out in the repair field. I always dreamed of the year 2000, and the flying cars, the robots that cleaned our houses, and all the rest of the futuristic ideas we had. Well, the year 2000 has come and gone, and no, we don't have flying cars, robots, or any of the stuff I dreamed about.

The automobile technology has gone forward a lot since I dreamed of these things. The fuel for this technology is OBDII and it is progressing with leaps and bounds. OBDII is pressing for the Zero Emissions vehicle, which is why we now have Hybrid vehicles. What are Hybrid vehicles? Hybrid electric vehicles (or HEVs) combine an internal combustion engine of a conventional vehicle with a battery and electric motor of an electric vehicle. This results in twice the fuel economy of conventional vehicles. This combination offers an extended range for the batteries. In the past, battery life was the major problem with electric vehicles. Constantly having to recharge the batteries kept the electric vehicles from hitting the main stream. The practical benefits of HEVs include improved fuel economy and lower emissions compared to conventional vehicles. The expected flexibility of HEVs will allow them to be used in a wide range of applications, from personal transportation to commercial hauling. In Europe, they have busses equipped with the HEV system. Imagine that, getting off a bus, and not being doused with the wave of diesel soot, as it pulls off. Where is the fun in that? Our kids will miss all the fun.

OBDIII?

Rumor has it the development of OBDIII is in process. This would take OBDII a step further by adding the technology of radio transponders to each vehicle. Most toll roads have an automated collection systems (Florida uses the Sun Pass), how much would it take to put transponders on vehicles so they would be able to report emissions problems (like the diagnostic codes, and VIN to identify the vehicle) directly to the local agency. Not much! This rather bothers me. Don't you think that would infringe on our privacy? Sounds like big brother to me.

Radio transponders are not that far fetched. Mercedes Benz has been using this technology for years. Mercedes calls this the Tele-Aid system. If a fault is detected, lets say the P0455 - EVAP Emission Control Sys. Gross leak detected. Not a serious fault, and the most common fault. The Tele-Aid system will notify Mercedes (in Montville, NJ), and a representative will call the customer on their Mercedes cell phone. The Rep will explain the procedure for tightening the gas cap, and have the customer start the vehicle. When the vehicle detects that the fault is not active, the vehicle will send a signal to the Rep, and in turn he will send the signal to the vehicle and clear the fault. Mercedes has another service. If the customer locks himself or herself out of the vehicle, the customer will call the Roadside Assistance, and a Mercedes Rep can send a signal to activate the door lock system to unlock the doors on the vehicle. Mercedes' idea is to eliminate the customer coming to the dealer for needless problems.

This is all pretty cool, huh? Mercedes is not the only ones that use this technology. On-Star has this technology for years also. On-Star has similar features.

Conclusion

As you can see, there are Millions of vehicles on the road with OBDII. Like it or not, OBDII has influenced your life, children's life, and hopefully their children's life with cleaner air and a cleaner environment.

Most mechanics think that OBDII is the government sticking their nose into the automotive market to make more money. Those mechanics are not that far off form the truth. BUT, in the long run, this is helping us with a cleaner environment, a more universal way to repair cars, and some really interesting technology. The automotive field has come a long way. And this is a good thing. For it sets the stage for the future. The Future? What a question. In the beginning of this chapter we discussed some of the vehicles that are in the future, and not so future. Only our mind is the limit to what we can do. So watch out automotive field, here comes the future, and OBDII is the fuel for it.

THE END

After a successful day of diagnosing OBDII, read the most important chapter!

Justina's Kitchen

ENTER

Treat your friends and family to a dinner; this time Venezuelan style.

As you buy each of the Secrets Revealed books, just cut out the recipes, collect them, and you will have... well, just that, a collection.

contributed by Justina David

My name is Justina David, I am from Maracaibo, Venezuela; my career in human resources has nothing to do with cooking, in fact I find it bizarre that I should be doing something like this, writing a recipe of one my mother's favorites dishes. I never thought of myself as a very good cook; of course, compared to my mom I am not even close.

Sadly, my mother passed away last year. She was a very outgoing person who loved to cook, dance and play cards (she was good at it, but she liked to cheat from time to time). She was a very cheerful, spontaneous person who always spoke her mind and was also a practical joker!

Curious about where the recipes originated? Christopher Columbus discovered the shores of Venezuela on his third voyage to the New World in 1498. It was the Italian geographer and navigator, Amerigo Vespucci (whose name also came to identify the New World), who named the area "Little Venice", or Venezuela, because of the resemblance of the native stilt houses found on the shores of Lake Maracaibo to the Venetian dwellings of the time. More importantly, some time between then and now, the dumpling was discovered and eventually moved into our kitchen.

The recipes on the following pages are dedicated to my Mami, Julia Elena.

Gracias.

Te quiero mucho y recuerdo siempre.

Life should be full of pleasures, and eating well is certainly one of them. Since you love cars, winning and winners, we are presenting a sure winner! When you prepare the following menu, everyone invited will be satisfied!

Afterwards, let us know, email us at obdii@kotzigpublishing.com. And, if you want to share your favorite winner menu with others, email it to the same address. You might want to include a paragraph or two about yourself, picture of your car, and whatever else you deem important. We might post it on the www.kotzigpublishing.com website, or even publish it in the next version of this book.

Venezuelan Menu
A sure winner!

[for party of 8]

Appetizer:

Cabbage salad

[Ensalada de Repollo]

Main Course:

**Beef filled dumplings
with salsa sauce
from the Maracaibo region**

[Bollos Pelones]

Dessert:

Baked Plantains with Honey

[Plátanos al horno con miel]

Cabbage Salad

First let's prepare the main ingredients:

- *Small green cabbage*
- *3 potatoes (+1 for the meat filling later)*
- *3 large carrots, shredded*
- *4 hard-boiled eggs*
- *1/2 bag of frozen green peas*

Slice the cabbage and put it in batches into boiling water for 1 minute. Remove and put in a large bowl. Leave the water boiling.

Cut 4 peeled potatoes into small, pea size pieces. Boil until soft, about 4 minutes (check if done before removing from water). Put 1/4 of the potatoes aside for the meat filling. Add the rest of the potatoes to the salad.

Hard-boil the eggs, about 10 minutes (you can use the same hot water used for potatoes). Cool, peel and cut the eggs into small pieces and add to the salad.

Shred the carrots and add to the cabbage.

Add the peas to the same bowl. Even though they are frozen, they will soften by the time you toss the salad with the dressing.

Now, for the dressing:

- *2 large spoons of mayonnaise*
- *3 tablespoons of your favorite oil*
- *2 teaspoons of vinegar*
- *ground pepper*
- *salt*
- *fresh or powdered garlic*

Mix all ingredients together well. Add to the salad, toss gently. Chill.

Beef filled dumplings with salsa sauce

Let's prepare the filling:

- *4 scallions*
- *1/2 green pepper*
- *1/2 red pepper*
- *1 onion*
- *1 large tomato*
- *3 teaspoons of minced garlic*
- *1 potato (from the salad recipe)*
- *1 bunch of cilantro, chopped*
- *3 tablespoons of olive oil*
- *3 tablespoons of Worcestershire sauce*
- *3 lbs of ground beef*
- *2 tablespoons of salt*

Cut all the scallions, peppers, onion, tomato and potato (unless already prepared at the time the salad was made) into small pieces.

Mince the garlic.

In a large pan, heat the oil on medium to high heat. Add the onion. One minute later add the Worcestershire sauce and all the scallions, peppers, onion, tomato and potato.

Cook for 5 minutes.
Add the meat and the salt. Stir occasionally until the meat is done, about 20 minutes.

Cover and let simmer for 15 more minutes.

Add all the cilantro.

Remove from the heat and let cool for 1 hour. If there is too much juice in the filling, strain it - otherwise it will be difficult to fill the dough.

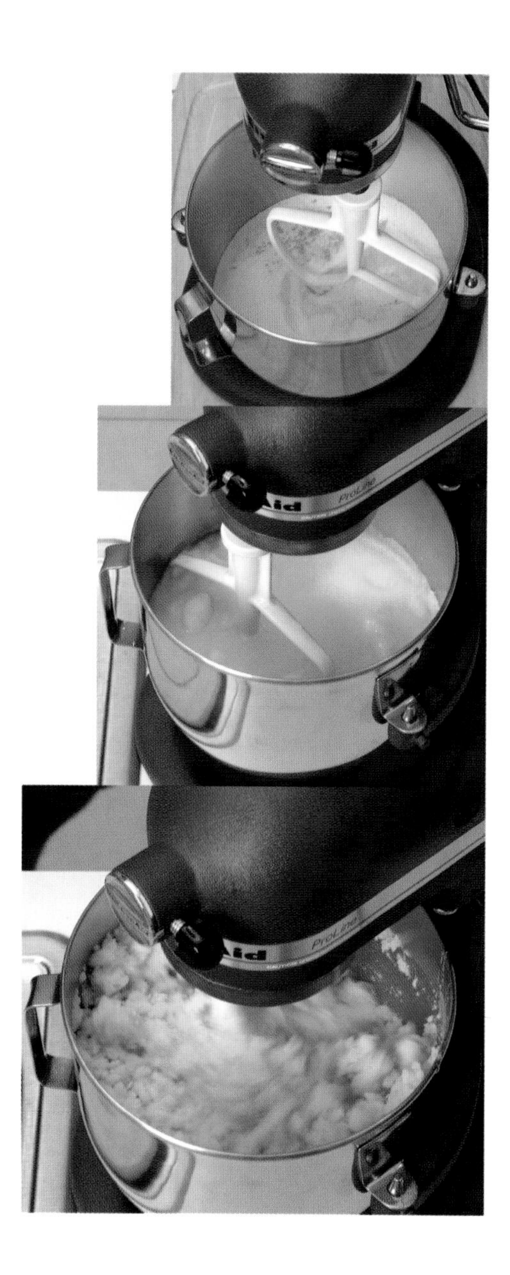

Now it is time for the corn dough:

- *1 bag of corn flour (1 kg / 2 lbs)*
- *1 teaspoon of saffron (or yellow coloring)*
- *9 cups of warm water*

Mix the dough. Wet your hands and form baseball sized balls (using the amounts above should make about 13 balls).

Using both hands, press your tumbs into the ball and form a bowl.

Fill each with 3 spoons of meat filling and close the dumpling into a ball shape again:

Boil in a generous amount of water for 10 minutes, 4 dumplings at a time.

We still need to make the salsa sauce:

- *1 green pepper*
- *1 red pepper*
- *2 small tomatoes*
- *1 onion*
- *3 scallions*
- *cilantro*
- *4 teaspoons of minced garlic*
- *2 tablespoons of olive oil*
- *2 tablespoons of Worcestershire sauce*
- *1 teaspoon of salt*
- *juice of 1 lemon*

In a pan, combine all the ingredients and simmer for 4 minutes.

Add 1/4 cup of water and simmer for 1 minute more.

Remove from the heat and add the lemon juice. Arrange one dumpling per serving on a large plate. Using two forks tear it open. Pour a generous amount of salsa sauce around it.

Serve with rice or bread.

Baked Plantains with Honey

- *4 ripe plantains*
- *4 tablespoons of honey*

Peel the plantains and cut lengthwise to create 8 halves. Place on buttered stainless sheet and bake at 375 ° Fahrenheit for 50 minutes. Remove the sheet from the oven, spoon on the honey and return to the oven for another 10 minutes.

Best if served right away. Yummy vanilla ice cream is optional.

POLAR

Generic OBDII Fault Codes (P0)

Basic checks for the P0xxx fault codes

If you notice that the majority of the faults have the same basic text. The only difference is the name of the component with the problem. If we were to list a fix for all the generic codes, we would need volumes of books. For instance, take this fault for GM: P0171, Fuel system lean bank 1. Since, GM has so many makes and models, there are 20 plus fixes for this one fault code. So, to make things simpler, here are some generic fixes for generic codes.

If you see the words:

Circuit Malfunction You need to check to see if there is a short, or open circuit to that component.

Circuit Range/Performance Problem Usually this will indicate that the component is the problem.

Circuit Low Input Usually this will indicate that the component is to be below its designated spec.

Circuit High Input Usually this will indicate that the component is to be above its designated spec.

Circuit Intermittent Usually this will indicate that the component has an intermittent problem. In most cases, a wiggle test is in order. Good practice is to hook up a scan tool, and check the Real Time Data of that component.

List of generic P0xxx fault codes (DTCs)

P0100 - Mass or Volume Air Flow Circuit Malfunction
P0101 - Mass or Volume Air Flow Circuit Range/Performance Problem
P0102 - Mass or Volume Air Flow Circuit Low Input
P0103 - Mass or Volume Air Flow Circuit High Input
P0104 - Mass or Volume Air Flow Circuit Intermittent
P0105 - Manifold Absolute Pressure/Barometric Pressure Circuit Malfunction
P0106 - Manifold Absolute Pressure/Barometric Pressure Circuit Range/Performance Problem
P0107 - Manifold Absolute Pressure/Barometric Pressure Circuit Low Input
P0108 - Manifold Absolute Pressure/Barometric Pressure Circuit High Input
P0109 - Manifold Absolute Pressure/Barometric Pressure Circuit Intermittent
P0110 - Intake Air Temperature Circuit Malfunction
P0111 - Intake Air Temperature Circuit Range/Performance Problem
P0112 - Intake Air Temperature Circuit Low Input
P0113 - Intake Air Temperature Circuit High Input
P0114 - Intake Air Temperature Circuit Intermittent
P0115 - Engine Coolant Temperature Circuit Malfunction
P0116 - Engine Coolant Temperature Circuit Range/Performance Problem
P0117 - Engine Coolant Temperature Circuit Low Input
P0118 - Engine Coolant Temperature Circuit High Input
P0119 - Engine Coolant Temperature Circuit Intermittent
P0120 - Throttle/Petal Position Sensor/Switch A Circuit Malfunction
P0121 - Throttle/Petal Position Sensor/Switch A Circuit Range/Performance Problem
P0122 - Throttle/Petal Position Sensor/Switch A Circuit Low Input
P0123 - Throttle/Petal Position Sensor/Switch A Circuit High Input
P0124 - Throttle/Petal Position Sensor/Switch A Circuit Intermittent
P0125 - Insufficient Coolant Temperature for Closed Loop Fuel Control

P0126 - Insufficient Coolant Temperature for Stable Operation
P0130 - O2 Sensor Circuit Malfunction (Bank 1 Sensor 1)
P0131 - O2 Sensor Circuit Low Voltage (Bank 1 Sensor 1)
P0132 - O2 Sensor Circuit High Voltage (Bank 1 Sensor 1)
P0133 - O2 Sensor Circuit Slow Response (Bank 1 Sensor 1)
P0134 - O2 Sensor Circuit No Activity Detected (Bank 1 Sensor 1)
P0135 - O2 Sensor Heater Circuit Malfunction (Bank 1 Sensor 1)
P0136 - O2 Sensor Circuit Malfunction (Bank 1 Sensor 2)
P0137 - O2 Sensor Circuit Low Voltage (Bank 1 Sensor 2)
P0138 - O2 Sensor Circuit High Voltage (Bank 1 Sensor 2)
P0139 - O2 Sensor Circuit Slow Response (Bank 1 Sensor 2)
P0140 - O2 Sensor Circuit No Activity Detected (Bank 1 Sensor 2)
P0141 - O2 Sensor Heater Circuit Malfunction (Bank 1 Sensor 2)
P0142 - O2 Sensor Circuit Malfunction (Bank 1 Sensor 3)
P0143 - O2 Sensor Circuit Low Voltage (Bank 1 Sensor 3)
P0144 - O2 Sensor Circuit High Voltage (Bank 1 Sensor 3)
P0145 - O2 Sensor Circuit Slow Response (Bank 1 Sensor 3)
P0146 - O2 Sensor Circuit No Activity Detected (Bank 1 Sensor 3)
P0147 - O2 Sensor Heater Circuit Malfunction (Bank 1 Sensor 3)
P0150 - O2 Sensor Circuit Malfunction (Bank 2 Sensor 1)
P0151 - O2 Sensor Circuit Low Voltage (Bank 2 Sensor 1)
P0152 - O2 Sensor Circuit High Voltage (Bank 2 Sensor 1)
P0153 - O2 Sensor Circuit Slow Response (Bank 2 Sensor 1)
P0154 - O2 Sensor Circuit No Activity Detected (Bank 2 Sensor 1)
P0155 - O2 Sensor Heater Circuit Malfunction (Bank 2 Sensor 1)
P0156 - O2 Sensor Circuit Malfunction (Bank 2 Sensor 2)
P0157 - O2 Sensor Circuit Low Voltage (Bank 2 Sensor 2)
P0158 - O2 Sensor Circuit High Voltage (Bank 2 Sensor 2)

P0159 - O2 Sensor Circuit Slow Response (Bank 2 Sensor 2)
P0160 - O2 Sensor Circuit No Activity Detected (Bank 2 Sensor 2)
P0161 - O2 Sensor Heater Circuit Malfunction (Bank 2 Sensor 2)
P0162 - O2 Sensor Circuit Malfunction (Bank 2 Sensor 3)
P0163 - O2 Sensor Circuit Low Voltage (Bank 2 Sensor 3)
P0164 - O2 Sensor Circuit High Voltage (Bank 2 Sensor 3)
P0165 - O2 Sensor Circuit Slow Response (Bank 2 Sensor 3)
P0166 - O2 Sensor Circuit No Activity Detected (Bank 2 Sensor 3)
P0167 - O2 Sensor Heater Circuit Malfunction (Bank 2 Sensor 3)
P0170 - Fuel Trim Malfunction (Bank 1)
P0171 - System too Lean (Bank 1)
P0172 - System too Rich (Bank 1)
P0173 - Fuel Trim Malfunction (Bank 2)
P0174 - System too Lean (Bank 2)
P0175 - System too Rich (Bank 2)
P0176 - Fuel Composition Sensor Circuit Malfunction
P0177 - Fuel Composition Sensor Circuit Range/Performance
P0178 - Fuel Composition Sensor Circuit Low Input
P0179 - Fuel Composition Sensor Circuit High Input
P0180 - Fuel Temperature Sensor A Circuit Malfunction
P0181 - Fuel Temperature Sensor A Circuit Range/Performance
P0182 - Fuel Temperature Sensor A Circuit Low Input
P0183 - Fuel Temperature Sensor A Circuit High Input
P0184 - Fuel Temperature Sensor A Circuit Intermittent
P0185 - Fuel Temperature Sensor B Circuit Malfunction
P0186 - Fuel Temperature Sensor B Circuit Range/Performance
P0187 - Fuel Temperature Sensor B Circuit Low Input
P0188 - Fuel Temperature Sensor B Circuit High Input

P0189 - Fuel Temperature Sensor B Circuit Intermittent
P0190 - Fuel Rail Pressure Sensor Circuit Malfunction
P0191 - Fuel Rail Pressure Sensor Circuit Range/Performance
P0192 - Fuel Rail Pressure Sensor Circuit Low Input
P0193 - Fuel Rail Pressure Sensor Circuit High Input
P0194 - Fuel Rail Pressure Sensor Circuit Intermittent
P0195 - Engine Oil Temperature Sensor Malfunction
P0196 - Engine Oil Temperature Sensor Range/Performance
P0197 - Engine Oil Temperature Sensor Low
P0198 - Engine Oil Temperature Sensor High
P0199 - Engine Oil Temperature Sensor Intermittent
P0200 - Injector Circuit Malfunction
P0201 - Injector Circuit Malfunction - Cylinder 1
P0202 - Injector Circuit Malfunction - Cylinder 2
P0203 - Injector Circuit Malfunction - Cylinder 3
P0204 - Injector Circuit Malfunction - Cylinder 4
P0205 - Injector Circuit Malfunction - Cylinder 5
P0206 - Injector Circuit Malfunction - Cylinder 6
P0207 - Injector Circuit Malfunction - Cylinder 7
P0208 - Injector Circuit Malfunction - Cylinder 8
P0209 - Injector Circuit Malfunction - Cylinder 9
P0210 - Injector Circuit Malfunction - Cylinder 10
P0211 - Injector Circuit Malfunction - Cylinder 11
P0212 - Injector Circuit Malfunction - Cylinder 12
P0213 - Cold Start Injector 1 Malfunction
P0214 - Cold Start Injector 2 Malfunction
P0215 - Engine Shutoff Solenoid Malfunction
P0216 - Injection Timing Control Circuit Malfunction

P0217 - Engine Over-temperature Condition
P0218 - Transmission Over Temperature Condition
P0219 - Engine Over-speed Condition
P0220 - Throttle/Petal Position Sensor/Switch B Circuit Malfunction
P0221 - Throttle/Petal Position Sensor/Switch B Circuit Range/Performance Problem
P0222 - Throttle/Petal Position Sensor/Switch B Circuit Low Input
P0223 - Throttle/Petal Position Sensor/Switch B Circuit High Input
P0224 - Throttle/Petal Position Sensor/Switch B Circuit Intermittent
P0225 - Throttle/Petal Position Sensor/Switch C Circuit Malfunction
P0226 - Throttle/Petal Position Sensor/Switch C Circuit Range/Performance Problem
P0227 - Throttle/Petal Position Sensor/Switch C Circuit Low Input
P0228 - Throttle/Petal Position Sensor/Switch C Circuit High Input
P0229 - Throttle/Petal Position Sensor/Switch C Circuit Intermittent
P0230 - Fuel Pump Primary Circuit Malfunction
P0231 - Fuel Pump Secondary Circuit Low
P0232 - Fuel Pump Secondary Circuit High
P0233 - Fuel Pump Secondary Circuit Intermittent
P0234 - Engine Overboost Condition
P0235 - Turbocharger Boost Sensor A Circuit Malfunction
P0236 - Turbocharger Boost Sensor A Circuit Range/Performance
P0237 - Turbocharger Boost Sensor A Circuit Low
P0238 - Turbocharger Boost Sensor A Circuit High
P0239 - Turbocharger Boost Sensor B Malfunction
P0240 - Turbocharger Boost Sensor B Circuit Range/Performance
P0241 - Turbocharger Boost Sensor B Circuit Low
P0242 - Turbocharger Boost Sensor B Circuit High
P0243 - Turbocharger Wastegate Solenoid A Malfunction
P0244 - Turbocharger Wastegate Solenoid A Range/Performance

P0245 - Turbocharger Wastegate Solenoid A Low
P0246 - Turbocharger Wastegate Solenoid A High
P0247 - Turbocharger Wastegate Solenoid B Malfunction
P0248 - Turbocharger Wastegate Solenoid B Range/Performance
P0249 - Turbocharger Wastegate Solenoid B Low
P0250 - Turbocharger Wastegate Solenoid B High
P0251 - Injection Pump Fuel Metering Control “A” Malfunction (Cam/Rotor/Injector)
P0252 - Injection Pump Fuel Metering Control “A” Range/Performance (Cam/Rotor/Injector)
P0253 - Injection Pump Fuel Metering Control “A” Low (Cam/Rotor/Injector)
P0254 - Injection Pump Fuel Metering Control “A” High (Cam/Rotor/Injector)
P0255 - Injection Pump Fuel Metering Control “A” Intermittent (Cam/Rotor/Injector)
P0256 - Injection Pump Fuel Metering Control “B” Malfunction (Cam/Rotor/Injector)
P0257 - Injection Pump Fuel Metering Control “B” Range/Performance (Cam/Rotor/Injector)
P0258 - Injection Pump Fuel Metering Control “B” Low (Cam/Rotor/Injector)
P0259 - Injection Pump Fuel Metering Control “B” High (Cam/Rotor/Injector)
P0260 - Injection Pump Fuel Metering Control “B” Intermittent (Cam/Rotor/Injector)
P0261 - Cylinder 1 Injector Circuit Low
P0262 - Cylinder 1 Injector Circuit High
P0263 - Cylinder 1 Contribution/Balance Fault
P0264 - Cylinder 2 Injector Circuit Low
P0265 - Cylinder 2 Injector Circuit High
P0266 - Cylinder 2 Contribution/Balance Fault
P0267 - Cylinder 3 Injector Circuit Low
P0268 - Cylinder 3 Injector Circuit High
P0269 - Cylinder 3 Contribution/Balance Fault
P0270 - Cylinder 4 Injector Circuit Low
P0271 - Cylinder 4 Injector Circuit High
P0272 - Cylinder 4 Contribution/Balance Fault

P0273 - Cylinder 5 Injector Circuit Low
P0274 - Cylinder 5 Injector Circuit High
P0275 - Cylinder 5 Contribution/Balance Fault
P0276 - Cylinder 6 Injector Circuit Low
P0277 - Cylinder 6 Injector Circuit High
P0278 - Cylinder 6 Contribution/Balance Fault
P0279 - Cylinder 7 Injector Circuit Low
P0280 - Cylinder 7 Injector Circuit High
P0281 - Cylinder 7 Contribution/Balance Fault
P0282 - Cylinder 8 Injector Circuit Low
P0283 - Cylinder 8 Injector Circuit High
P0284 - Cylinder 8 Contribution/Balance Fault
P0285 - Cylinder 9 Injector Circuit Low
P0286 - Cylinder 9 Injector Circuit High
P0287 - Cylinder 9 Contribution/Balance Fault
P0288 - Cylinder 10 Injector Circuit Low
P0289 - Cylinder 10 Injector Circuit High
P0290 - Cylinder 10 Contribution/Balance Fault
P0291 - Cylinder 11 Injector Circuit Low
P0292 - Cylinder 11 Injector Circuit High
P0293 - Cylinder 11 Contribution/Balance Fault
P0294 - Cylinder 12 Injector Circuit Low
P0295 - Cylinder 12 Injector Circuit High
P0296 - Cylinder 12 Contribution/Range Fault
P0300 - Random/Multiple Cylinder Misfire Detected
P0301 - Cylinder 1 Misfire Detected
P0302 - Cylinder 2 Misfire Detected
P0303 - Cylinder 3 Misfire Detected

P0304 - Cylinder 4 Misfire Detected
P0305 - Cylinder 5 Misfire Detected
P0306 - Cylinder 6 Misfire Detected
P0307 - Cylinder 7 Misfire Detected
P0308 - Cylinder 8 Misfire Detected
P0309 - Cylinder 9 Misfire Detected
P0311 - Cylinder 11 Misfire Detected
P0312 - Cylinder 12 Misfire Detected
P0320 - Ignition/Distributor Engine Speed Input Circuit Malfunction
P0321 - Ignition/Distributor Engine Speed Input Circuit Range/Performance
P0322 - Ignition/Distributor Engine Speed Input Circuit No Signal
P0323 - Ignition/Distributor Engine Speed Input Circuit Intermittent
P0325 - Knock Sensor 1 Circuit Malfunction (Bank 1 or Single Sensor)
P0326 - Knock Sensor 1 Circuit Range/Performance (Bank 1 or Single Sensor)
P0327 - Knock Sensor 1 Circuit Low Input (Bank 1 or Single Sensor)
P0328 - Knock Sensor 1 Circuit High Input (Bank 1 or Single Sensor)
P0329 - Knock Sensor 1 Circuit Intermittent (Bank 1 or Single Sensor)
P0330 - Knock Sensor 2 Circuit Malfunction (Bank 2)
P0331 - Knock Sensor 2 Circuit Range/Performance (Bank 2)
P0332 - Knock Sensor 2 Circuit Low Input (Bank 2)
P0333 - Knock Sensor 2 Circuit High Input (Bank 2)
P0334 - Knock Sensor 2 Circuit Intermittent (Bank 2)
P0335 - Crankshaft Position Sensor A Circuit Malfunction
P0336 - Crankshaft Position Sensor A Circuit Range/Performance
P0337 - Crankshaft Position Sensor A Circuit Low Input
P0338 - Crankshaft Position Sensor A Circuit High Input
P0339 - Crankshaft Position Sensor A Circuit Intermittent
P0340 - Camshaft Position Sensor Circuit Malfunction

P0341 - Camshaft Position Sensor Circuit Range/Performance
P0342 - Camshaft Position Sensor Circuit Low Input
P0343 - Camshaft Position Sensor Circuit High Input
P0344 - Camshaft Position Sensor Circuit Intermittent
P0350 - Ignition Coil Primary/Secondary Circuit Malfunction
P0351 - Ignition Coil A Primary/Secondary Circuit Malfunction
P0352 - Ignition Coil B Primary/Secondary Circuit Malfunction
P0353 - Ignition Coil C Primary/Secondary Circuit Malfunction
P0354 - Ignition Coil D Primary/Secondary Circuit Malfunction
P0355 - Ignition Coil E Primary/Secondary Circuit Malfunction
P0356 - Ignition Coil F Primary/Secondary Circuit Malfunction
P0357 - Ignition Coil G Primary/Secondary Circuit Malfunction
P0358 - Ignition Coil H Primary/Secondary Circuit Malfunction
P0359 - Ignition Coil I Primary/Secondary Circuit Malfunction
P0360 - Ignition Coil J Primary/Secondary Circuit Malfunction
P0361 - Ignition Coil K Primary/Secondary Circuit Malfunction
P0362 - Ignition Coil L Primary/Secondary Circuit Malfunction
P0370 - Timing Reference High Resolution Signal A Malfunction
P0371 - Timing Reference High Resolution Signal A Too Many Pulses
P0372 - Timing Reference High Resolution Signal A Too Few Pulses
P0373 - Timing Reference High Resolution Signal A Intermittent/Erratic Pulses
P0374 - Timing Reference High Resolution Signal A No Pulses
P0375 - Timing Reference High Resolution Signal B Malfunction
P0376 - Timing Reference High Resolution Signal B Too Many Pulses
P0377 - Timing Reference High Resolution Signal B Too Few Pulses
P0378 - Timing Reference High Resolution Signal B Intermittent/Erratic Pulses
P0379 - Timing Reference High Resolution Signal B No Pulses
P0380 - Glow Plug/Heater Circuit "A" Malfunction

P0381 - Glow Plug/Heater Indicator Circuit Malfunction
P0382 - Glow Plug/Heater Circuit "B" Malfunction
P0385 - Crankshaft Position Sensor B Circuit Malfunction
P0386 - Crankshaft Position Sensor B Circuit Range/Performance
P0387 - Crankshaft Position Sensor B Circuit Low Input
P0388 - Crankshaft Position Sensor B Circuit High Input
P0389 - Crankshaft Position Sensor B Circuit Intermittent
P0400 - Exhaust Gas Recirculation Flow Malfunction
P0401 - Exhaust Gas Recirculation Flow Insufficient Detected
P0402 - Exhaust Gas Recirculation Flow Excessive Detected
P0403 - Exhaust Gas Recirculation Circuit Malfunction
P0404 - Exhaust Gas Recirculation Circuit Range/Performance
P0405 - Exhaust Gas Recirculation Sensor A Circuit Low
P0406 - Exhaust Gas Recirculation Sensor A Circuit High
P0407 - Exhaust Gas Recirculation Sensor B Circuit Low
P0408 - Exhaust Gas Recirculation Sensor B Circuit High
P0410 - Secondary Air Injection System Malfunction
P0411 - Secondary Air Injection System Incorrect Flow Detected
P0412 - Secondary Air Injection System Switching Valve A Circuit Malfunction
P0413 - Secondary Air Injection System Switching Valve A Circuit Open
P0414 - Secondary Air Injection System Switching Valve A Circuit Shorted
P0415 - Secondary Air Injection System Switching Valve B Circuit Malfunction
P0416 - Secondary Air Injection System Switching Valve B Circuit Open
P0417 - Secondary Air Injection System Switching Valve B Circuit Shorted
P0418 - Secondary Air Injection System Relay "A" Circuit Malfunction
P0419 - Secondary Air Injection System Relay "B" Circuit Malfunction
P0420 - Catalyst System Efficiency Below Threshold (Bank 1)
P0421 - Warm Up Catalyst Efficiency Below Threshold (Bank 1)

P0422 - Main Catalyst Efficiency Below Threshold (Bank 1)
P0423 - Heated Catalyst Efficiency Below Threshold (Bank 1)
P0424 - Heated Catalyst Temperature Below Threshold (Bank 1)
P0430 - Catalyst System Efficiency Below Threshold (Bank 2)
P0431 - Warm Up Catalyst Efficiency Below Threshold (Bank 2)
P0432 - Main Catalyst Efficiency Below Threshold (Bank 2)
P0433 - Heated Catalyst Efficiency Below Threshold (Bank 2)
P0434 - Heated Catalyst Temperature Below Threshold (Bank 2)
P0440 - Evaporative Emission Control System Malfunction
P0441 - Evaporative Emission Control System Incorrect Purge Flow
P0442 - Evaporative Emission Control System Leak Detected (small leak)
P0443 - Evaporative Emission Control System Purge Control Valve Circuit Malfunction
P0444 - Evaporative Emission Control System Purge Control Valve Circuit Open
P0445 - Evaporative Emission Control System Purge Control Valve Circuit Shorted
P0446 - Evaporative Emission Control System Vent Control Circuit Malfunction
P0447 - Evaporative Emission Control System Vent Control Circuit Open
P0448 - Evaporative Emission Control System Vent Control Circuit Shorted
P0449 - Evaporative Emission Control System Vent Valve/Solenoid Circuit Malfunction
P0450 - Evaporative Emission Control System Pressure Sensor Malfunction
P0451 - Evaporative Emission Control System Pressure Sensor Range/Performance
P0452 - Evaporative Emission Control System Pressure Sensor Low Input
P0453 - Evaporative Emission Control System Pressure Sensor High Input
P0454 - Evaporative Emission Control System Pressure Sensor Intermittent
P0455 - Evaporative Emission Control System Leak Detected (gross leak)
P0460 - Fuel Level Sensor Circuit Malfunction
P0461 - Fuel Level Sensor Circuit Range/Performance
P0462 - Fuel Level Sensor Circuit Low Input
P0463 - Fuel Level Sensor Circuit High Input

P0464 - Fuel Level Sensor Circuit Intermittent
P0465 - Purge Flow Sensor Circuit Malfunction
P0466 - Purge Flow Sensor Circuit Range/Performance
P0467 - Purge Flow Sensor Circuit Low Input
P0468 - Purge Flow Sensor Circuit High Input
P0469 - Purge Flow Sensor Circuit Intermittent
P0470 - Exhaust Pressure Sensor Malfunction
P0471 - Exhaust Pressure Sensor Range/Performance
P0472 - Exhaust Pressure Sensor Low
P0473 - Exhaust Pressure Sensor High
P0474 - Exhaust Pressure Sensor Intermittent
P0475 - Exhaust Pressure Control Valve Malfunction
P0476 - Exhaust Pressure Control Valve Range/Performance
P0477 - Exhaust Pressure Control Valve Low
P0478 - Exhaust Pressure Control Valve High
P0479 - Exhaust Pressure Control Valve Intermittent
P0480 - Cooling Fan 1 Control Circuit Malfunction
P0481 - Cooling Fan 2 Control Circuit Malfunction
P0482 - Cooling Fan 3 Control Circuit Malfunction
P0483 - Cooling Fan Rationality Check Malfunction
P0484 - Cooling Fan Circuit Over Current
P0485 - Cooling Fan Power/Ground Circuit Malfunction
P0500 - Vehicle Speed Sensor Malfunction
P0501 - Vehicle Speed Sensor Range/Performance
P0502 - Vehicle Speed Sensor Low Input
P0503 - Vehicle Speed Sensor Intermittent/Erratic/High
P0505 - Idle Control System Malfunction
P0506 - Idle Control System RPM Lower Than Expected

P0507 - Idle Control System RPM Higher Than Expected
P0510 - Closed Throttle Position Switch Malfunction
P0520 - Engine Oil Pressure Sensor/Switch Circuit Malfunction
P0521 - Engine Oil Pressure Sensor/Switch Circuit Range/Performance
P0522 - Engine Oil Pressure Sensor/Switch Circuit Low Voltage
P0523 - Engine Oil Pressure Sensor/Switch Circuit High Voltage
P0530 - A/C Refrigerant Pressure Sensor Circuit Malfunction
P0531 - A/C Refrigerant Pressure Sensor Circuit Range/Performance
P0532 - A/C Refrigerant Pressure Sensor Circuit Low Input
P0533 - A/C Refrigerant Pressure Sensor Circuit High Input
P0534 - Air Conditioner Refrigerant Charge Loss
P0550 - Power Steering Pressure Sensor Circuit Malfunction
P0551 - Power Steering Pressure Sensor Circuit Range/Performance
P0552 - Power Steering Pressure Sensor Circuit Low Input
P0553 - Power Steering Pressure Sensor Circuit High Input
P0554 - Power Steering Pressure Sensor Circuit Intermittent
P0560 - System Voltage Malfunction
P0561 - System Voltage Unstable
P0562 - System Voltage Low
P0563 - System Voltage High
P0565 - Cruise Control On Signal Malfunction
P0566 - Cruise Control Off Signal Malfunction
P0567 - Cruise Control Resume Signal Malfunction
P0568 - Cruise Control Set Signal Malfunction
P0569 - Cruise Control Coast Signal Malfunction
P0570 - Cruise Control Accel Signal Malfunction
P0571 - Cruise Control/Brake Switch A Circuit Malfunction
P0572 - Cruise Control/Brake Switch A Circuit Low

P0573 - Cruise Control/Brake Switch A Circuit High
P0574 - Cruise Control Related Malfunction
P0575 - Cruise Control Related Malfunction
P0576 - Cruise Control Related Malfunction
P0577 - Cruise Control Related Malfunction
P0578 - Cruise Control Related Malfunction
P0579 - Cruise Control Related Malfunction
P0580 - Cruise Control Related Malfunction
P0600 - Serial Communication Link Malfunction
P0601 - Internal Control Module Memory Check Sum Error
P0602 - Control Module Programming Error
P0603 - Internal Control Module Keep Alive Memory (KAM) Error
P0604 - Internal Control Module Random Access Memory (RAM) Error
P0605 - Internal Control Module Read Only Memory (ROM) Error
P0606 - PCM Processor Fault
P0608 - Control Module VSS Output "A" Malfunction
P0609 - Control Module VSS Output "B" Malfunction
P0620 - Generator Control Circuit Malfunction
P0621 - Generator Lamp "L" Control Circuit Malfunction
P0622 - Generator Field "F" Control Circuit Malfunction
P0650 - Malfunction Indicator Lamp (MIL) Control Circuit Malfunction
P0654 - Engine RPM Output Circuit Malfunction
P0655 - Engine Hot Lamp Output Control Circuit Malfunction
P0656 - Fuel Level Output Circuit Malfunction
P0700 - Transmission Control System Malfunction
P0701 - Transmission Control System Range/Performance
P0702 - Transmission Control System Electrical
P0703 - Torque Converter/Brake Switch B Circuit Malfunction

P0704 - Clutch Switch Input Circuit Malfunction
P0705 - Transmission Range Sensor Circuit malfunction (PRNDL Input)
P0706 - Transmission Range Sensor Circuit Range/Performance
P0707 - Transmission Range Sensor Circuit Low Input
P0708 - Transmission Range Sensor Circuit High Input
P0709 - Transmission Range Sensor Circuit Intermittent
P0710 - Transmission Fluid Temperature Sensor Circuit Malfunction
P0711 - Transmission Fluid Temperature Sensor Circuit Range/Performance
P0712 - Transmission Fluid Temperature Sensor Circuit Low Input
P0713 - Transmission Fluid Temperature Sensor Circuit High Input
P0714 - Transmission Fluid Temperature Sensor Circuit Intermittent
P0715 - Input/Turbine Speed Sensor Circuit Malfunction
P0716 - Input/Turbine Speed Sensor Circuit Range/Performance
P0717 - Input/Turbine Speed Sensor Circuit No Signal
P0718 - Input/Turbine Speed Sensor Circuit Intermittent
P0719 - Torque Converter/Brake Switch B Circuit Low
P0720 - Output Speed Sensor Circuit Malfunction
P0721 - Output Speed Sensor Range/Performance
P0722 - Output Speed Sensor No Signal
P0723 - Output Speed Sensor Intermittent
P0724 - Torque Converter/Brake Switch B Circuit High
P0725 - Engine Speed input Circuit Malfunction
P0726 - Engine Speed Input Circuit Range/Performance
P0727 - Engine Speed Input Circuit No Signal
P0728 - Engine Speed Input Circuit Intermittent
P0730 - Incorrect Gear Ratio
P0731 - Gear 1 Incorrect ratio
P0732 - Gear 2 Incorrect ratio

P0733 - Gear 3 Incorrect ratio
P0734 - Gear 4 Incorrect ratio
P0735 - Gear 5 Incorrect ratio
P0736 - Reverse incorrect gear ratio
P0740 - Torque Converter Clutch Circuit Malfunction
P0741 - Torque Converter Clutch Circuit Performance or Stuck Off
P0742 - Torque Converter Clutch Circuit Stuck On
P0743 - Torque Converter Clutch Circuit Electrical
P0744 - Torque Converter Clutch Circuit Intermittent
P0745 - Pressure Control Solenoid Malfunction
P0746 - Pressure Control Solenoid Performance or Stuck Off
P0747 - Pressure Control Solenoid Stuck On
P0748 - Pressure Control Solenoid Electrical
P0749 - Pressure Control Solenoid Intermittent
P0750 - Shift Solenoid A Malfunction
P0751 - Shift Solenoid A Performance or Stuck Off
P0752 - Shift Solenoid A Stuck On
P0753 - Shift Solenoid A Electrical
P0754 - Shift Solenoid A Intermittent
P0755 - Shift Solenoid B Malfunction
P0756 - Shift Solenoid B Performance or Stuck Off
P0757 - Shift Solenoid B Stuck On
P0758 - Shift Solenoid B Electrical
P0759 - Shift Solenoid B Intermittent
P0760 - Shift Solenoid C Malfunction
P0761 - Shift Solenoid C Performance or Stuck Off
P0762 - Shift Solenoid C Stuck On
P0763 - Shift Solenoid C Electrical

P0764 - Shift Solenoid C Intermittent
P0765 - Shift Solenoid D Malfunction
P0766 - Shift Solenoid D Performance or Stuck Off
P0767 - Shift Solenoid D Stuck On
P0768 - Shift Solenoid D Electrical
P0769 - Shift Solenoid D Intermittent
P0770 - Shift Solenoid E Malfunction
P0771 - Shift Solenoid E Performance or Stuck Off
P0772 - Shift Solenoid E Stuck On
P0773 - Shift Solenoid E Electrical
P0774 - Shift Solenoid E Intermittent
P0780 - Shift Malfunction
P0781 - 1-2 Shift Malfunction
P0782 - 2-3 Shift Malfunction
P0783 - 3-4 Shift Malfunction
P0784 - 4-5 Shift Malfunction
P0785 - Shift/Timing Solenoid Malfunction
P0786 - Shift/Timing Solenoid Range/Performance
P0787 - Shift/Timing Solenoid Low
P0788 - Shift/Timing Solenoid High
P0789 - Shift/Timing Solenoid Intermittent
P0790 - Normal/Performance Switch Circuit Malfunction
P0801 - Reverse Inhibit Control Circuit Malfunction
P0803 - 1-4 Upshift (Skip Shift) Solenoid Control Circuit Malfunction
P0804 - 1-4 Upshift (Skip Shift) Lamp Control Circuit Malfunction

Appendix B: OBDII Acronyms

3-2TS - 3-2 Timing Solenoid

3GR - Third Gear

4GR - Fourth Gear

4WD - Four Wheel Drive

A

A/C - Air Conditioning

A/F - Air Fuel Ratio

A/T - Automatic Transaxle

A/T - Automatic Transmission

AAT - Ambient Air Temperature

ACL - Air Cleaner

AFC - Air Flow Control

AIR - Secondary Air Injection

AP - Accelerator Pedal

APP - Accelerator Pedal Position

B

B+ - Battery Positive Voltage

BARO - Barometric Pressure

BC - Blower Control

BPP - Brake Pedal Position

BUS N - Bus Negative

BUS P - Bus Positive

C

CAC - Charge Air Cooler

CAN - Controller Area Network

CARB - Carburetor

CC - Climate Control

CCS - Coast Clutch Solenoid

CFI - Continuous Fuel Injection

CFV - Critical Flow Venturi

CKP - Crankshaft Position

CL - Closed Loop

CMP - Camshaft Position

CO - Carbon Monoxide

CO2 - Carbon Dioxide

CPP - Clutch Pedal Position

CTOX - Continuous Trap oxidizer

CTP - Closed Throttle Position

CVS - Constant Volume Sampler

D

DFI - Direct Fuel Injection

DI - Distributor Ignition

DLC - Data Link Connector

DM - Drive Motor

DMCM - Drive Motor Control Module

DMCT - Drive Motor Coolant Temperature

DMPI - Drive Motor Power Inverter

DTC - Diagnostic Trouble Code

DTM - Diagnostic Test Mode

E

EC - Engine Control

ECL - Engine Coolant Level

ECM - Engine Control Module

ECT - Engine Coolant Temperature

EEC - Electronic Engine Control

EEPROM - Electrically Erasable Programmable Read Only Memory

EFE - Early Fuel Evaporation

EFI - Electronic Fuel Injection

EFT - Engine Fuel Temperature

EGR - Exhaust Gas Recirculation

EGRT - Exhaust Gas Recirculation Temperature

EI - Electronic Ignition

EM - Engine Modification

EMR - Electronic Module Retard

EOP - Engine Oil Pressure

EOT - Engine Oil Temperature

EP - Exhaust Pressure

EPROM - Erasable Programmable Read Only Memory

ESC - Electronic Spark Control

EST - Electronic Spark Timing

EVAP - Evaporative Emission

F

F4WD - Full Time Four Wheel Drive

FC - Fan Control

FEEPROM - Flash Electrically Erasable Programmable Read Only Memory

FEPROM - Flash Erasable Programmable Read Only Memory

FF - Flexible Fuel

FID - Flame Ionization Detector

FLI - Fuel Level Indicator

FP - Fuel Pump

FRP - Fuel Rail Pressure

FRT - Fuel Rail Temperature

FT - Fuel Trim

FTP - Fuel Tank Pressure

FTT - Fuel Tank Temperature

FWD - Front Wheel Drive

G

GCM - Governor Control Module

GEN - Generator

GND - Ground

GPM - Grams Per Mile

H

HC - Hydrocarbon

HCDS - High Clutch Drum Speed

HEI - High Energy Ignition

HO2S - Heated Oxygen Sensor
HPC - High Pressure Cutoff

I

I/M - Inspection and Maintenance
IA - Intake Air
IAC - Idle Air Control
IAT - Intake Air Temperature
IC - Ignition Control
ICM - Ignition Control Module
ICP - Injection Control Pressure
IFI - Indirect Fuel Injection
IFS - Inertia Fuel Shutoff
IMRC - Intake Manifold Runner Control
IMT - Intake Manifold Tuning
ISC - Idle Speed Control
ISS - Input Shaft Speed

K

KS - Knock Sensor

L

LOAD - Calculated Load Value

M

MAF - Mass Airflow
MAP - Manifold Absolute Pressure
MAT - Manifold Air Temperature
MC - Mixture Control
MDP - Manifold Differential Pressure
MFG - Manufacturer
MFI - Multiport Fuel Injection
MIL - Malfunction Indicator Lamp
MST - Manifold Surface Temperature
MVZ - Manifold Vacuum Zone

N

NDIR - Non-Dispersive Infra Red
NOx - Nitrogen Oxides
NVRAM - Non-Volatile Random Access Memory

O

O2 - Oxygen
O2S - Oxygen Sensor
OBD - On Board Diagnostic
OC - Oxidation Catalytic Converter
ODS - Overdrive Drum Speed
OL - Open Loop
OSS - Output Shaft Speed

P

PAIR - Pulsed Secondary Air Injection
PC - Pressure Control
PCM - Power Train Control Module
PCV - Positive Crankcase Ventilation
PID - Parameter Identification

PNP - Park/Neutral Position
PR - Pressure Relief
PROM - Programmable Read Only Memory
PSC - Power Steering Control
PSP - Power Steering Pressure
PTC - Pending Trouble Code
PTOX - Periodic Trap Oxidizer
PWM - Pulse Width Modulation

R

RAM - Random Access Memory
RM - Relay Module
ROM - Read Only Memory
RPM - Engine Speed
RWD - Rear Wheel Drive

S

S4WD - Selectable Four Wheel Drive
SC - Supercharger
SCB - Supercharger Bypass
SES - Service Engine Soon dash light, now referred to as MIL
SFI - Sequential Multiport Fuel Injection
SPL - Smoke Puff Limiter
SR - Service Reminder Indicator
SRT - System Readiness Test
SS - Shift Solenoid
ST - Scan Tool

T

TAC - Throttle Actuator Control
TB - Throttle Body
TBI - Throttle Body Fuel Injection
TC - Turbocharger
TCC - Torque Converter Clutch
TCCP - Torque Converter Clutch Pressure
TCM - Transmission Control Module
TE - Thermal Expansion
TFP - Transmission Fluid Pressure
TFT - Transmission Fluid Temperature
TP - Throttle Position
TPI - Tuned Port Injection
TPS - Throttle Position Sensor
TR - Transmission Range
TRLHP - Track Road Load Horsepower
TSS - Turbine Shaft Speed
TVV - Thermal Vacuum Valve
TWC - Three Way Catalytic Converter
TWC+OC - Three Way + Oxidation Catalytic Converter

V

VAC - Vacuum
VAF - Volume Airflow
VCM - Vehicle Control Module

VCRM - Variable Control Relay Module

VIN - Vehicle Identification Number

VR - Voltage Regulator

VSS - Vehicle Speed Sensor

W

WOT - Wide Open Throttle

WU-OC - Warm-up Oxidation Catalytic Converter

WU-TWC - Warm-up Three Way Catalytic Converter

C

CAC - Charge Air Cooler

CAN - Controller Area Network

CARB - Carburetor

CC - Climate Control

CCS - Coast Clutch Solenoid

CFI - Continuous Fuel Injection

CFV - Critical Flow Venturi

CKP - Crankshaft Position

CL - Closed Loop

CMP - Camshaft Position

CO - Carbon Monoxide

CO2 - Carbon Dioxide

CPP - Clutch Pedal Position

CTOX - Continuous Trap oxidizer

CTP - Closed Throttle Position

CVS - Constant Volume Sampler

D

DFI - Direct Fuel Injection

DI - Distributor Ignition

DLC - Data Link Connector

DM - Drive Motor

DMCM - Drive Motor Control Module

DMCT - Drive Motor Coolant Temperature

DMPI - Drive Motor Power Inverter

DTC - Diagnostic Trouble Code

DTM - Diagnostic Test Mode

E

EC - Engine Control

ECL - Engine Coolant Level

ECM - Engine Control Module

ECT - Engine Coolant Temperature

EEC - Electronic Engine Control

EEPROM - Electrically Erasable Programmable Read Only Memory

EFE - Early Fuel Evaporation

EFI - Electronic Fuel Injection

EFT - Engine Fuel Temperature

EGR - Exhaust Gas Recirculation

EGRT - Exhaust Gas Recirculation Temperature

Appendix C: OBDII Terms and Definitions

Accelerator Pedal
A foot operated device, which, directly or indirectly, controls the flow of fuel and/or air to the engine, controlling engine speed.

Accumulator
A vessel in which liquid or gas is stored, usually at greater than atmospheric pressure

Actuator
A mechanism for moving or controlling something indirectly instead of by hand.
Compare: Solenoid, Relay and Valve

Air Conditioning
A vehicular accessory system that modifies the passenger compartment air by cooling and drying the air.

Alternator
See Generator

Battery
An electrical storage device designed to produce a DC voltage by means of an electrochemical reaction

Blower
A device designed to supply a current of air at a moderate pressure. A blower usually consists of an impeller assembly, a motor and a suitable case. The blower case is usually designed as part of a ventilation system.
Compare: Fan

Brake
A device for retarding motion, usually by means of friction

Body
The assembly of components, windows, doors, seats, etc; that provides enclosures for passengers and/or cargo in a motor vehicle. It may or may not include the hood and fenders.

Bypass
Providing a secondary path to relieve pressure in the primary passage.

Camshaft
A shaft on which phased cams are mounted. The camshaft is used to regulate the opening and closing of the intake and exhaust valves.

Canister
An evaporative emission canister contains activated charcoal, which absorbs fuel vapors and holds them until the vapors can be purged at an appropriate time.

Capacitor
An electrical device for accumulating and holding a charge of electricity.

Carbon Dioxide
A heavy, colorless gas that can be found as a product of complete combustion.

Carbon Monoxide
A colorless, odorless gas that can be found as a product of incomplete combustion.

Carburetor
A mechanism which automatically mixes fuel with air in the proper proportions to provide a desired power output from a spark ignition internal combustion engine.

Catalyst
A substance that can increase or decrease the rate of a chemical reaction between substances without being consumed in the process.

Chassis
The suspension, steering and braking elements of a vehicle.

Circuit
A complete electrical path or channel; usually includes the source of electric energy. Circuit may also describe the electrical path between two or more components. May also be used with fluids, air or liquid.

Cleaner
A device used in the intake system of parts that require clean air. An air cleaner usually has a filter in it to trap particulates and only pass clean air through.

Climate
The temperature/ventilation in the passenger compartment.

Closed Loop
An operating condition or mode that enables modification of programmed instructions based on a feedback system.

Clutch
A mechanical device, which uses mechanical, magnetic, or friction type connections to facilitate engaging or disengaging of two shafts or rotating members.

Code
A system of symbols (as letters, numbers, or words) used to represent meaning of information.

Coil (Ignition)
A device consisting of windings of conductors around an iron core, designed to increase the voltage, and for use in a spark ignition system.

Control
A means or a device to direct and regulate a process or guide the operation of a machine, apparatus, or system.

Converter (Catalytic)
An in-line exhaust system device used to reduce the level of engine exhaust emissions.

Converter (Torque)
A device, which by its design multiplies the torque in a fluid coupling between an engine and transmission/transaxle.

Coolant
A fluid used for heat transfer. Coolants usually contain additive such as rust inhibitors and antifreeze.

Cooler
A heat exchanger that reduces the temperature of the named medium.

Crankshaft
The part of an engine, which converts the reciprocating motion of the pistons to rotary motion.

Data
General term for information, usually represented by numbers, letters and/or symbols.

Device
A piece of equipment or a mechanism designed for a specific purpose or function. DO NOT use “Device” as a Base Word.

Diagnostics
The process of identifying the cause or nature of a condition, situation, or problem, in order to determine corrective action in repair of automotive systems.

Differential
A device with an arrangement of gears designed to permit the division of power to two shafts.

Distributor
A mechanical device designed to switch a high voltage secondary circuit from an ignition coil to spark plugs in the proper firing sequence.

Drive
A device which provides a fixed increase or decrease ratio of relative rotation between its input and output shafts.

Driver
A switched electronic device that controls output state.

Electrical
A type of device or system using resistors, motors, generators, incandescent lamps, switches, capacitors, batteries, inductors, or wires.
Compare: Electronic.

Electronic
A type of device or system using solid state devices or thermionic elements such as diodes, transistors, integrated circuits, vacuum fluorescent displays, and liquid crystal displays; or the storage, retrieval, and display of information through media such as magnetic tape, laser disc, electronic read-only memory (ROM), and random access memory (RAM).

Engine
A machine designed to convert thermal energy into mechanical energy to produce force or motion.

Exhaust
Gaseous byproducts of combustion emitted from an engine.

Fan
A device designed to supply a current of air. A fan may also have a frame, motor, wiring, harness, and the like.
Compare: Blower

Fuel
Any combustible substance burned to provide heat or power. Typical fuels include gasoline and diesel fuel. Other types of fuel include ethanol, methanol, natural gas, propane, or a combination.

Generator
A rotating machine designed to convert mechanical energy into electrical energy.

Glow Plug
A combustion chamber heat-generating device to aid starting diesel engines.

Governor
A device designed to automatically limit engine speed.

Ground
An electrical conductor used as a common return for an electric circuit(s) and with a relative zero potential.

Hydrocarbon
An organic compound containing various carbon and hydrogen molecules, which occur in fuel.

Idle
Rotational speed of an engine with vehicle at rest and accelerator pedal not depressed.

Ignition
System used to provide high voltage spark for internal combustion engines.

Indicator
A device, which visually presents vehicle condition information transmitted or relayed from some other source.

Injector
A device for delivering metered, pressurized fuel to the intake system or the cylinders.

Input Shaft
A shaft in a device that is “driven” by the previous element in the power train.

Intake Air
Air drawn through a cleaner and distributed to each cylinder for use in combustion.

Inverter
A device, which converts direct current to alternating current.

Knock (Engine)
The sharp, metallic sound produced when two pressure fronts collide in the combustion chamber of an engine.

Level
The magnitude of a quantity considered in relation to an arbitrary reference value.

Line
A generic service term used to describe a system of pipes, tubes and hoses

Link (Electrical/Electronic)
General term used to indicate the existence of communication facilities between two points.

Manifold
A device designed to collect or distribute fluid, air, or the like.
Compare: Rail

Memory
A device in which data can be stored and used when needed.

Mode
One of several alternative conditions or methods of operating a device or control module.

Module (Electrical/Electronic)
A self-contained group of electrical/electronic components, which is designed as a single replaceable unit.

Motor
A machine that converts kinetic energy, such as electricity, into mechanical energy.
Compare: Actuator.

Nitrogen Oxides
Various combinations of nitrogen and oxygen atoms, which can be a product of incomplete combustion.

Open Loop
An operating condition or mode based on programmed instructions and not modified by a feedback system.

Output Shaft
A shaft in a device that drives the next element in the power train.

Oxygen
A colorless, tasteless, odorless gas that supports combustion.

Park/Neutral
The selected non-drive modes of the transmission.

Power Steering
A system, which provides additional force to the steering mechanism, reducing the driver's steering effort.

Powertrain
The elements of a vehicle by which motive power is generated and transmitted to the driven axles.

Pressure
Unless otherwise noted, is gauge pressure.

Pressure (Absolute)
The pressure referenced to a perfect vacuum.

Pressure (Atmospheric)
The pressure of the surrounding air at any given temperature and altitude. Sometimes called barometric pressure.

Pressure (Barometric)
Pertaining to atmospheric pressure or the results obtained by using a barometer.

Pressure (Differential)
The pressure difference between two regions, such as between the intake manifold and the atmospheric pressures.

Pressure (Gauge)
The amount by which the total absolute pressure exceeds the ambient atmospheric pressure.

Pump
A device used to raise, transfer or compress fluids by suction, pressure or both.

Radiator
A radiator is a liquid to air heat transfer device having a tank(s) and core(s) specifically designed to reduce the temperature of the coolant in an internal combustion engine cooling system.

Rail
A manifold for fuel injection fuel.
Compare: Manifold.

Refrigerant
A substance used as a heat transfer agent in an air conditioning system.

Relay
A generally electromechanical device in which connections in one circuit are opened or closed by changes in another circuit.
Compare: Actuator, Solenoid, Switch.

Regulator (Voltage)
A device that automatically controls the functional output of another device by adjusting the voltage to meet a specified value.

Scan Tool
A device that interfaces with and communicates information on a data link.

Secondary Air
Air provided to the exhaust system.

Sensor
The generic name for a device that senses either the absolute value or a change in a physical quantity such as temperature, pressure, rotation or flow rate, and converts that change into an electrical quantity signal.
Compare: Transducer.

Shift Solenoid
A device that controls shifting in an automatic transmission.

Signal (Electrical/Electronic)
A fluctuating electric quantity, such as voltage or current, whose variations represent information.

Solenoid
A device consisting of an electrical coil which when energized, produces a magnetic field in a plunger, which is pulled to a central position. A solenoid may be used as an actuator in a valve or switch.
Compare: Actuator, Relay, Switch

Solid State
Crystalline circuit structure used to perform electronic functions. Examples of such structures include transistors, diodes, integrated circuits and other semiconductors.

Speed
The magnitude of velocity (regardless of direction).

Supercharger
A mechanically driven device that pressurizes the intake air, thereby increasing the density of charge air and the consequent power output from a given engine displacement.

Switch
A device for making, breaking or changing the connections in an electrical circuit.
Compare: Relay, Solenoid, Valve

System
A group of interacting mechanical or electrical components serving a common purpose.

Tank
A storage device for liquid or gas.

Test
A procedure whereby the performance of a product is measured under various conditions.

Thermal Expansion
The expansion of a solid, liquid, or gas due to a change in temperature.

Throttle
A valve for regulating the supply of a fluid, usually air or a fuel/air mix, to an engine.

Transaxle
A device consisting of a transmission and axle drive gears assembled in the same case.
Compare: Transmission.

Transducer
A device that receives energy from one system and re-transmits (transfers) it, often in a different form, to another system. For example, the cruise control transducer converts a vehicle speed signal to a modulated vacuum output to control a servo.
Compare: Sensor

Transmission
A device which selectively increases or decreases the ration of relative rotation between its input and output shafts.
Compare: Transaxle.

Troubleshooting
See Diagnostics.

Turbine Shaft
A shaft in a device that is driven by a turbine.

Turbocharger
A centrifugal device driven by exhaust gases that pressurize the intake air, thereby increasing the density of charge air and the consequent power output from a given engine displacement.

Ultraviolet
The portion of the electromagnetic spectrum between violet visible light and x-rays.

Vacuum
A circuit in which pressure has been reduced below the ambient atmospheric pressure.

Valve
A device by which the flow of a liquid, gas, vacuum, or loose material in bulk may be started.

Vapor
A substance in its gaseous state as distinguished from the liquid or solid state.

Volatile
Vaporized at normal temperatures; or not permanent.

Wastegate
A valve to limit charge air pressure by allowing exhaust gases to bypass the turbocharger.

Wheel
A circular frame of hard material that may be solid, partially solid, or spoked and capable of turning on an axle.

NOTES

NOTES

the secrets have been revealed